# THE STUBBORNLY PERSISTENT

## Melting The Frozen River

## Of Spacetime

**Anderthal Kord**

ISBN-13: 978-1985830844

# CONTENTS

# INTRODUCTION

Einstein summarized the problem before us. "People like us, who believe in physics," he declared, "know that the distinction between the past, present and future is only a stubbornly persistent illusion."

An illusion? Really?

Our problem is that we do believe in physics, and we admire Einstein, but we also believe there are very real distinctions between the past, the present and the future.

And we are going to remain stubbornly persistent in our belief.

At stake is the meaning of existence—the meaning of the existence of physical reality, and the meaning of *us*—our lives, our struggles and our dreams.

If there is no distinction between the past, the present and the future, then a flowing present moment is not transforming an open future potentiality into an actualized past. Rather, all times equally exist, and we exist equally within those times—our past still exists, our future already exists, and our present *simply* exists, having no particular status or relevance. The present moment is not, as we believe, the time of our life—the time of being actively aware, the time of meaning and purpose, of reacting spontaneously to situations, of making decisions and plans that will affect our lives and our being. No, the present is merely an arbitrary dividing line between the past and the future, within a timeless, unchanging reality.

What does the existence of physical reality even mean, if there is no flow of time, no true movement, no evolution or change? What does the existence of human reality even mean, if there is no distinction between our past, our present and our future?

All of the normal mental human activities we engage in every day, such as focusing on the world around us, in the present, seeking out rewards, threats and opportunities; contemplating an unknown future, so that we might prepare for what is to come; mulling over the memories of our past, so that we might take pride in our accomplishments, and learn from our mistakes; thinking, in the present, about what to say, how to act, who to be—deciding what moral philosophy we should have, what ethical decisions to make—these are all just strange and futile wastes of time and effort, within this frozen river of time. All

# INTRODUCTION

moments simply, eternally exist, and we have no hope of changing a single whit of what is, and what is to be.

But we humans, to a large degree, identify with our mental activity—we think of our ongoing mental dialogue as "who" we are. So where does this leave us living, breathing human beings? What meaning can we give to our hopes and our plans, our triumphs and our failures, if we simply exist within time?

We live in the present. This is the moment of our awareness, our being, of reacting and responding to the world, living our life. In the present, we give meaning to the past, as the source of our memories and experiences. The past may be over with and done, out of reach, but it still has an importance to our lives and our identities. And the future has an equal importance, to our present, as the source of possibility, the land of the unknown, full of potential and surprises—and of hope, that our actions of today may bring about a better tomorrow.

Our present, then, is not some arbitrary dividing line between the past and the future, but the moment of meaning and actuality, of movement and change—our present is forever flowing into the future and leaving the past behind.

Our lives revolve around these three different modes of time; they form the basis of our actions, our thoughts, our concepts of purpose and meaning. If the distinction between the past, the present and the future is only a stubbornly persistent illusion, then so are the very lives we live.

# THE STUBBORNLY PERSISTENT

Einstein was stating an inevitable conclusion of his theory of relativity. Relativity unifies space with time into a single physical edifice, called the fourth-dimensional spacetime continuum. Within this universe, all of space, and all of time, all of physical reality, simply exists. There is no flowing present within this world, only an eternally unchanging, frozen block of moments.

Within the fourth-dimensional spacetime continuum, the flowing river of time is a frozen river of spacetime. Fourth-dimensionally, there is no movement, no change, no growth or evolution. There is human awareness, but that awareness is frozen with the rest of reality. It cannot influence events, choose between outcomes, transform our surroundings, or our selves. Within this reality, we have no responsibility for our actions, we have no control over any outcome or event—we have no free will.

What is the point of being consciously aware, if we are merely helpless spectators observing a predetermined existence? What meaning can we give to our own personal identities, our sense of self, if we are not the product of our reactions and decisions, the sum of our choices and beliefs?

We stubbornly persistent simply cannot accept this reality, as is.

We are not mute witnesses to reality, but active participants. Our actions matter. *We* matter. This is the reality *we choose* to live in.

We choose a reality that has free will.

# INTRODUCTION

There is another, more subtle problem with this reality. If all moments in time have an equal existence—if there is no distinction between the past, the present and the future—then the Big Bang is not the "origin" of time and space, it is not "the reason" physical reality exists—it is simply a boundary marking off physical existence, in the same sense that the tip of a baseball bat is a boundary of the baseball bat, and not the "origin" or "the reason" the baseball bat exists. All of physical reality simply, eternally exists, without any apparent cause or origin.

So, where did we actually come from? How are we here?

Within relativity's spacetime, these questions are meaningless. We did not come from anywhere, we simply *are*, eternally existing, without any origin, without any cause, and without any further explanation.

To summarize our problem, then, the past, present and future are arbitrary and illusory divisions of time, within relativity's fourth-dimensional spacetime continuum, while in the human continuum, the division of time into the past, the present and the future define our lives, our identities and our existence.

So, what do we do?

Do we choose, perhaps, our humanity, and simply ignore our science? Or do we choose our science, and deny our humanity?

# THE STUBBORNLY PERSISTENT

No, we choose neither of these alternatives; we choose to keep our humanity *and* our science.

Our human minds, and our natural curiosity, our search for patterns and meanings created science in the first place, and with our scientific progress we have steadily and relentlessly transformed the world. For science to then turn around and deny the power of our human minds to affect the world, to claim we are merely spectators, simply seems wrong, if not actually perverse. But we still cannot use this as an excuse to deny our science, because science has become our sixth sense, our clairvoyance, our extrasensory perception that allows us to peer ever deeper into the mysteries of the universe.

We choose to keep our science, and with it relativity's spacetime.

We just ask it to agree with human experience.

Which means that, somewhere along the way, we will simply have to reach out, grab spacetime by its full head of hair and then drag it, kicking and screaming, into the land of free will.

Or, perhaps, we might simply give it a little nudge, the slightest of tweaks. Whatever works.

Welcome to the world of the stubbornly persistent.

# CHAPTOR ONE

# This Strange World

When a lifetime friend of Einstein's, Michele Besso, died in 1955, Einstein wrote to the family, "Now he has departed from this strange world a little ahead of me. That signifies nothing. For those of us who believe in physics, the distinction between the past, present and future is only a stubbornly persistent illusion."

Einstein followed his friend a few weeks later.

When he wrote these words, Einstein had many years behind him in which to accept the implications of living in a fourth-dimensional spacetime continuum. His theory of relativity had done away with the concept of "simultaneous events," and in so doing had transformed the flowing river of time into a frozen block universe.

# THE STUBBORNLY PERSISTENT

Simultaneous events are all of those events that are happening at the same moment of time. At any one moment—say, "right now"—there are a great number of things happening, and, according to Newtonian physics, everyone will agree on what those things are. Einstein showed that this is not true; if two observers (two "inertial systems") are moving in different directions, or at different speeds, they will not agree on which things are happening at a given moment. One person will say two events happened at the same time, while the other person will say one of those events happened *before* (or after) the other event. Although this may seem like a minor quibble, its effects are far reaching. This is because each person's viewpoint is equally valid, and there are countless numbers of viewpoints in the world, with everyone traveling in different directions and different speeds. And if everyone's viewpoint is equally valid, and they all have different interpretations of which events are in the past, the present and the future, the inevitable conclusion is that all of these moments must be equally real, and they are merely being viewed differently by different observers. And the discrepancies become ever more pronounced the farther away the observers are from each other, and the faster they are traveling, so that the entire universe ultimately becomes crisscrossed with a practically infinite number of definitions of past, present and future.

Let us say I define certain events as happening

"right now." Because they are in my present moment, I have no trouble believing they exist— indeed, this is essentially our definition of "existence." These events exist. But you, traveling by me (you are one of those events existing in my "now") experience some of these events in your past, and some other of these events you experience in your future (when you come to see them, in your future). Now, your viewpoint is as correct as mine, and yet, the events you see as "past" and "future" I see as "present." Logically, then, the events are real—they exist— because I see them as in the present, even though you see them as existing in the past or future. And the events *you* see as happening in your present, I will see as in my past and future. But because they are in *your* present, those events "exist."

The obvious, inevitable conclusion is that all moments in time must equally exist. And so, the distinction between the past, the present and the future is only a stubbornly persistent illusion.

In our houses, in our cars and airplanes, in our immediate neighborhoods, these effects are miniscule—but they are there. And the universe is far larger than our immediate neighborhoods, and it is the universe we are concerned with, here—the past, the present and the future of the universe itself.

In the 1954 edition of his book, *Relativity, the Special and General Theory*, Einstein worded these ideas a little more technically:

"In the first place we must guard against the

opinion that the four-dimensionality of reality has been newly introduced for the first time by this theory. Even in classical physics the event is localized by four numbers, three spatial co-ordinates and a time co-ordinate; the totality of physical "events" is thus thought of as being embedded in a four-dimensional continuous manifold. But on the basis of classical mechanics this four-dimensional continuum breaks up objectively into the one-dimensional time and into three-dimensional spatial sections, only the latter of which contain simultaneous events. This resolution is the same for all inertial systems. The simultaneity of two definite events with reference to one inertial system involves the simultaneity of these events in reference to all inertial systems. This is what is meant when we say that the time of classical mechanics is absolute. According to the special theory of relativity it is otherwise. The sum total of events which are simultaneous with a selected event exist, it is true, in relation to a particular inertial system, but no longer independently of the choice of the inertial system. The four-dimensional continuum is now no longer resolvable objectively into sections, all of which contain simultaneous events; "now" loses for the spatially extended world its objective meaning. It is because of this that space and time must be regarded as a four-dimensional continuum that is objectively unresolvable."

There is, in other words, no moment "now"

that exists throughout the universe. The events I define as happening "now" will be past and future events to everyone else. The events exist, in time, but a single "time" does not exist—only all of time.

Einstein concludes, "Since there exist in this four-dimensional structure no longer any sections which represent "now" objectively, the concepts of happening and becoming are indeed not completely suspended, but yet complicated. It appears therefore more natural to think of physical reality as a four-dimensional existence, instead of, as hitherto, the *evolution* of a three-dimensional existence."

Three-dimensional space is not evolving through time; rather, all of space and time simply exist as a single, complete and unchanging fourth-dimensional spacetime continuum.

This may be a lot to take in, all at once, so let us hear other wordings of the subject from various experts. In Bertrand Russell's book, *The ABC of Relativity*, he explains: "We may now recapitulate the reasons which have made it necessary to substitute "spacetime" for space and time. The old separation of space and time rested upon the belief that there was no ambiguity in saying that two events in distant places happened at the same time; consequently, it was thought that we could describe the topography of the universe at a given instant in purely spatial terms. But now that simultaneity has become relative to a particular observer, this is no longer possible. What is, for one observer, a description of the state of the world at a given instant is, for another observer, a series of events at various different times, whose relations are not merely spatial but also temporal. For the same reason, we are concerned with *events*, rather than with bodies… When we know the time and place of an event in one observer's system of reckoning, we can calculate its time and place according to another observer. But we must know the time as well as the place, because we can no longer ask what is its place for the new observer at the "same" time as for the old observer. There is no such thing as the "same" time for different observers, unless they are at rest relatively to each other. We need four measurements to fix a position, and four measurements fix the position of an event in space-time, not merely of a body in space. Three measurements are not enough to fix any position. That is the essence of what is meant by the substitution of space-time for space and time."

According to the mathematics and geometry of relativity, when we think of the universe as existing "right now, at this moment," this is only an illusion we create in our minds. Things exist, but they exist within spacetime, not in any special moment that we can physically define as unique.

In 1905, Einstein formulated his special theory of relativity; in 1915, he had finished his general theory of relativity. In between these two events, Hermann Minkowski entered the scene and altered the course Einstein was on, as Adam Frank elaborates in his book, *About Time*:

"Hermann Minkowski, a German mathematician and physicist, was known for casting physics problems into the language of geometry, the language of spatial relationship. In reviewing Einstein's early papers, Minkowski saw a way to translate relativity into a powerful geometric vocabulary that would alter all future descriptions of cosmology. Relativity, he discovered, was not simply concerned with objects extended in space (the traditional study of geometry); instead, it described the structure of *events* in space and time taken as a whole.

"Events were the real objects of concern in relativity... Minkowski recognized that what mattered was not the location of these events in three-dimensional space alone or their location in time alone. Instead, relativity provided relationships between a cosmic web of events in something much larger. Minkowski cast relativity into the geometry of space-time, a new four-dimensional reality. Space-time was the new stage on which the drama of physics would be enacted...

"The philosophical implications of the new perspective were startling... The future and past took on a different character in the so-called block universe of space-time. In this vision of relativity, next Tuesday, which we consider to be the future, already exists. The past and future are reduced to events that exist together in the totality of a timeless, eternal block of space-time."

# THE STUBBORNLY PERSISTENT

In his book, *Cycles of Time*, Roger Penrose explains:

"Minkowski's space-time has a different kind of geometric structure, giving a curious twist to Euclid's ancient idea of geometry. It provides an *overall* geometry to space-time, making it one indivisible whole, which completely encodes the structure of Einstein's special relativity."

Penrose goes on to describe how two people walking past each other, here on earth, would, because of their relative motion, each define the present moment on our nearby galaxy, Andromeda, different by several weeks.

The faster any two observers separate, and the larger the distances involved, the more out of sync in time the universe becomes.

In fact, reversing this description, two aliens on a planet in Andromeda, walking past one another, would define the present moment, here on earth, different by several weeks. To one alien you are where you are right now, while to the other alien you are where you were a few weeks ago, or perhaps where you will be a few weeks from now. And yet, to each alien, this is their *present* moment.

The point is that any and all observers in the universe have an equally valid definition of what exists in their past, their present and their future. If *you* were traveling by earth right now, at a fraction of the speed of light, *you* would define the universe differently than everyone else on earth. And *you* could potentially be traveling in any direction, at any speed.

All viewpoints are equally valid, equally real.

In his book, *The Fabric of the Cosmos*, Brian Greene compares spacetime to a loaf of bread: "...the slices making up the loaf are the nows of a given observer; each slice represents space at one moment of time from his or her perspective. The union obtained by placing slice next to slice, in the order in which the observer experiences them, fills out a region of spacetime. If we take this perspective to a logical extreme and imagine that each slice depicts *all* of space at a given moment of time according to one observer's viewpoint, and if we include every possible slice, from the ancient past to the distant future, the loaf will encompass all of the universe throughout all time— the whole of spacetime. Every occurrence, regardless of when or where, is represented by some point in the loaf.

"...there is convincing evidence that the spacetime loaf—the totality of spacetime, not slice by single slice—is real. A less than widely appreciated implication of Einstein's work is that special relativistic reality treats all times equally. Although the notion of *now* plays a central role in our worldview, relativity subverts our intuition once again and declares ours an egalitarian universe in which every moment is as real as any other."

Every moment is as real as any other. The

distinction between the past, present and future is only a stubbornly persistent illusion.

In our human reality, the past is over, gone, and the future is yet to be. In between these two is the present, the moment we live in and experience, the moment of awareness and free will. More, our present moment is continuously moving forward through time, forever flowing into the future and leaving the past behind—*carrying us with it.*

None of this has any meaning whatsoever in relativity's fourth-dimensional spacetime continuum. In spacetime there is no flowing present, because there is no flow. There is only the entire loaf of spacetime.

Julian Barbour tells us, in *The End of Time,* "Hermann Minkowski's ideas have penetrated deep into the psyche of modern physicists. They find it hard to contemplate any alternative to his grand vision...

"...the resulting space-time structure, now called Minkowski space-time...is more like a loaf of bread, through which you can slice in any way...

"There is no natural way in which time can flow in Minkowski's space-time. At least within classical physics, space-time is a block—it simply is. This is known as the block universe view of time. Everything—past, present and future—is there at once."

In *The Matter Myth*, Paul Davies and John Gribbin word it like this: "This unified four-dimensional spacetime description has proved highly successful in explaining many physical phenomena, and is now the accepted view of the physical world. Powerful though it is, it has removed from the picture any vestige of a personal "now," or the division of time into past, present and future... The reason for this is that, according to relativity theory, time does not "happen" bit by bit, or moment by moment: it is stretched out, like space, in its entirety. Time is simply "there"."

And there we are, stuck in time, believing (for some unexplainable reason) that time flows, that we move relentlessly forward through time, stumbling into the future with each new moment and each new day, that life is change and growth, possibility and surprises, that our actions have consequences we are responsible for—which is the reason we have law and order, rewards and punishment—that we have the ability to shape our futures and mold our surroundings and change things for the better—that we have free will.

But this is all an illusion. As David Deutsch sums up in his book, *The Fabric of Reality*,

"So there is no single 'present moment,' except subjectively. From the point of view of an observer at a particular moment, that moment is indeed singled out, and may uniquely be called 'now' by that observer, just as any position in space is singled out as 'here' from the point of view of an observer at that position. But objectively, no moment is privileged as being more 'now' than the others, just as no position is privileged as being more 'here' than other positions. The subjective 'here' may move through space, as the observer moves. Does the subjective 'now' likewise move through time? ...Certainly not. Even subjectively, 'now' *does not move through time.* It is often said that the present *seems* to be moving forwards in time because the present is defined only relative to our consciousness, and our consciousness is sweeping forwards through the moments. But our consciousness does not, and could not, do that. When we say that our consciousness 'seems' to pass from one moment to the next we are merely paraphrasing the commonsense theory of the flow of time. But it makes no more sense to think of a single 'moment of which we are conscious' moving from one moment to another than it does to think of a single present moment, or anything else, doing so. *Nothing* can move from one moment to another. To exist at all at a particular moment means to exist there for ever. Our consciousness exists at *all* our (waking) moments."

The question we have to ask, is, why does the illusion of a flowing time even exist? Where did it come from, and how could it have come from anything, if there is only the state of 'existing eternally as is'? Why, in other words, does the illusion of a flowing time eternally exist within a static, unchanging spacetime?

David Deutsch accepts this situation, but he

also believes in a far richer reality—a quantum reality. He believes we exist in a multiverse, a reality of infinitely branching universes. In the quantum multiverse we are surrounded by infinite copies and variations of ourselves and our world. In this reality, the very concept of time dissolves away, because those other universes *include* what we view as past and future times.

Other times *are* other universes.

This is a bizarre reality, strangely appealing from a philosophically scientific point of view. It is not, however, the one we will be following here.

Our goal, the goal of our book, is to have *our* reality, *our* individual lives have meaning, a meaning that emerges naturally within a flowing river of time, with true distinctions between the past, the present and the future. To achieve this, we will do whatever is necessary—we may even have to become diabolical and devious, at some point, twisting scientific theories inside out, perhaps, or flipping them upside down, confiscating some and discarding others with reckless scientific abandon. We may even appropriate those other universes of David Deutsch for our own personal gain—as if we are engaging in a sort of evolutionary battle for the survival of the most stubbornly persistent.

Here is a little experiment: Imagine an awareness existing at an instantaneous moment of time. How to do this? As a first step, you probably must first figure out if awareness needs the passage of time in order to even exist—but the true point is this: as you think these thoughts, as you ponder these ideas, you will no doubt notice moments in time passing—which essentially answers the riddle.

This suggests to us that there is something—*something*—lacking in relativity's description of reality.

Let us now, therefore, begin focusing our gaze on spacetime's full head of hair...

# CHAPTOR TWO

# The Stubbornly Persistent

Einstein's theory of relativity is not the only theory in physics that conflicts with our human perception of a flowing time. In his book, *The Emperor's New Mind*, Roger Penrose describes how our personal experience of time's flow is left out of the equations of Newton, Maxwell, Schrödinger, and all the other highly successful descriptions of our physical universe. Time, within physics, is symmetrical between the past and the future, with no mention of, or need for, a flowing present moment.

Penrose then turns his attention to relativity's unavoidable conclusion that all of spacetime must be "definite," without uncertainty, and with no flow of time. The future is not being transformed into the past by an ever flowing present; there is only a fixed and immutable fourth-dimensional spacetime continuum.

But Penrose does not accept this: "It seems to me that there are severe discrepancies between what we consciously feel, concerning the flow of time, and what our (marvelously accurate) theories assert about the reality of the physical world. These discrepancies must surely be telling us something deep about the physics that presumably must actually underlie our conscious perceptions—assuming (as I believe) that what underlies these perceptions can indeed be understood in relation to some appropriate kind of physics. At least it seems to be clearly the case that whatever physics is operating, it must have an essentially time-asymmetrical ingredient, i.e. it must make a distinction between the past and the future."

Roger Penrose is a famous mathematical physicist, a philosopher of science—and, evidently, he, too, is stubbornly persistent.

In his book, *About Time*, Paul Davies summarizes the lack of physical evidence for a flowing time: "The passage of time is often viewed as the progress of "the now" *through* time. We can envisage the time dimension stretched out as a line of fate, and a particular instant—"now"—being singled out as a little glowing point. As "time goes on," so the light moves steadily up the time line towards the future. Needless to say, physicists can find nothing of this in the objective world: no little light, no privileged present, no migration up the time line."

# THE STUBBORNLY PERSISTENT

Davies is one of the physicists he refers to, but he goes on to admit,

"Yet, as a human being, I find it impossible to relinquish the sensation of a flowing time and a moving present moment. It is something so basic to my experience of the world that I am repelled by the claim that it is only an illusion or misperception. It seems to me there is an aspect of time of great significance that we have so far overlooked in our description of the physical universe."

Paul Davies—even though he has departed from this strange world—still eternally exists within the fabric of spacetime—as one of the stubbornly persistent.

These are physicists grappling with what their science tells them versus how they actually perceive reality. In his book, *Genius—The Life and Science of Richard Feynman*, James Gleick summarizes their plight: "The physicist drawing his diagrams obtains a God's-eye view. In the space-time picture a line representing the path of a particle through time simply exists, past and future visible together. The four-dimensional space-time manifold displays all eternity at once.

"The laws of nature are not rules controlling the metamorphosis of what is into what will be. They are descriptions of patterns that exist, all at once, in the whole tapestry. The picture is hard to reconcile with our everyday sense that time is special. Even the physicist has his memories of the past and his aspirations for the future, and no spacetime diagram quite obliterates the difference between them."

Lee Smolin is a theoretical physicist who has written an entire book on the reality of human time, as opposed to the timelessness of physical laws. He begins his book, *Time Reborn*, with the question, What is time?

"This deceptively simple question is the single most important problem facing science as we probe more deeply into the fundamentals of the universe. All of the mysteries physicists and cosmologists face—from the Big Bang to the future of the universe, from the puzzles of quantum physics to the unification of the forces and particles—come down to the nature of time."

He describes the conflict between human experience and timeless existence: "Time is the most pervasive aspect of our everyday experience. Everything we think, feel, or do reminds us of its existence. We perceive the world as a flow of moments that make up our life. But physicists and philosophers alike have long told us (and many people think) that time is the ultimate illusion."

For example, he says, "Relativity strongly suggests that the whole history of the world is a timeless unity; present, past and future have no meaning apart from human subjectivity. Time is just another dimension of space, and the sense we have of experiencing moments passing is an illusion behind which is a timeless reality."

But, Smolin believes, "as strongly as one can believe anything in science… Time will turn out to be the only aspect of our everyday experience that *is* fundamental. The fact that it is always some moment in our perception, and that we experience that moment as one of a flow of moments, is not an illusion. It is the best clue we have to fundamental reality."

We have been hearing from serious and dedicated scientists who are also explorers and adventurers testing the limits of human knowledge. They believe in the spirit and progress of science, while also believing that our human perception of reality is an important clue that something fundamental is missing from our present scientific theories.

There are many more out there, in space and in time; we will collectively refer to them all as The League of the Stubbornly Persistent. They follow their own unique paths, searching for answers to this ultimate mystery.

We will also follow our own unique path, one that will be quite different from theirs. In one sense, our approach will be rasher and more reckless than theirs, lacking in a certain scientific rigor they might frown upon—at some point, we will offer an actual solution to these mysteries, a wild guess, perhaps, or a desperate, conjuring incantation. We will also shamelessly borrow from present theories, past theories and the speculations of others, searching for a way to see what is presently known from a brand-new perspective—a perspective that will hopefully "make sense" to ordinary human beings.

In another sense, however, all we are offering, here, is a simple thought experiment. Is it possible to imagine a reality in which relativity's fourth-dimensional spacetime continuum has free will embedded within its structure? How would this reality look? How would it operate?

In order to do this, we admit, we may have to make a wild and crazy assumption, or two, in order for all the rest to fall into place. Our justification for this is simple: we are going to do whatever it takes to unify relativity's spacetime with our human perception of a flowing time, which means we are going to remain stubbornly persistent, to the very end.

Before we begin, however, we have one more reality to explore.

# CHAPTOR THREE

## A Stranger World

How do we know that spacetime exists—that it has an actual, physical existence—and that it is not simply an idea, or a mental construct, that exists only in the mind?

We know this because things fall to the ground when they are dropped, because the moon circles the world day after day, and because the sun shines bright and continuously. These things occur because of gravity, and it is the *curvature* of spacetime that produces what we perceive as the force of gravity.

In order for spacetime to curve, it must be a real, physical entity. Otherwise, what would be curving? Mathematical symbols? Mental constructs?

General relativity describes the curvature of spacetime, and it does so with very precise, and complex, mathematical equations. The curvature of spacetime *is* gravity, and gravity rules our lives, our world and the universe at large. General relativity, then, describes the physical reality we live in, and that reality is spacetime. Space and time united in a single reality that *curves* to produce gravity.

There is, however, an embarrassing omission to this reality: it does not include quantum mechanics.

Along with gravity, there are three forces ruling the universe: the electromagnetic force, the weak nuclear force and the strong nuclear force. These three forces, plus gravity, are responsible for all physical interactions in the world around us.

Quantum mechanics is the basis for these other three forces. Therefore, we will now say a few words about quantum mechanics.

Weird. Strange. Bizarre. Crazy.

We could say a few more words, but instead let us hear how Richard Feynman compares relativity with quantum mechanics: "There was a time when the newspapers said that only twelve men understood the theory of relativity. I do not believe there ever was such a time. There might have been a time when only one man did because he was the only guy who caught on, before he wrote his paper. But after people read the paper a lot of people understood the theory of relativity in one way or other, certainly more than twelve. On the other hand I think I can safely say that nobody understands quantum mechanics."

This is because quantum mechanics is simply weird, strange, bizarre and crazy.

For example, it is impossible to measure a quantum particle's position and its momentum at the same time. Why not? That is just the way it is. This measurement is easily performed in our ordinary world, but it is impossible to do with quantum particles. There is even an equation telling us precisely how impossible this is: the equation tells us that the more accurately we measure a particle's position, the *less* we can know about the particle's momentum, or, conversely, the more accurately we measure a particle's momentum, the less we can know about the particle's position. This strange rule—the uncertainty principle—is built into the fabric of reality of the very small. Nobody knows why.

In his book, *The Black Hole War,* Leonard Susskind explains, "The Uncertainty Principle was the great divide that separated physics into the prequantum *classical* era and the postmodern era of quantum "weirdness." Classical physics consists of everything that came before Quantum Mechanics, including Newton's theory of motion, Maxwell's theory of light, and Einstein's theories of relativity. Classical physics is deterministic; quantum physics is full of uncertainty.

"The Uncertainty Principle is a strange

and audacious claim that was made in 1927 by the twenty- six- year-old Werner Heisenberg shortly after he and Erwin Schrödinger discovered the mathematics of Quantum Mechanics. Even in an era of many unfamiliar ideas, it stood out as especially bizarre. Heisenberg made no claim that there was a limitation on how accurately one could measure the position of an object. The coordinates that locate a particle in space can be determined to any desired degree of precision. He also put no limitation on how accurately the velocity of an object could be measured. What he claimed was that no experiment, however complex or ingenious, could ever be designed to measure both position and velocity simultaneously."

And no experiment has ever been designed to do this. Werner Heisenberg had crystallized into a principle what other scientists were trying to make sense of with their experiments and mathematics.

As David Lindley describes in his book, *Uncertainty*:

"Pauli and Dirac had seen that there was something strange about the way quantum physics manifested itself to the outside world. Heisenberg's uncertainty pinned down that strangeness, put a number on it, and—perhaps most important to Heisenberg—dashed any lingering hope that Schrödinger with his waves could restore some sort of classical reality to physics.

"...Formally, Schrödinger's equation is deterministic in the old-fashioned sense. That is, if you know the wave function for some system at a certain time, you can calculate it exactly and unambiguously at any later time—provided, that is, you don't attempt any observation in the interim. Measurement is what causes Born's probability interpretation of the wave to swing into action: different results are possible, with different likelihoods.

"Heisenberg's uncertainty nailed down the inescapability of the discord between one possible measurement and another. An observer can choose to measure this, that, or the other, but has to put up with resulting incommensurabilities. And that uncertainty feeds into the future development of the system. The quantum wave function changes to reflect the fact that one particular measurement outcome occurred and other possibilities didn't—and that in turn influences the possible outcomes of subsequent measurements that might be made."

What you find in the quantum world depends upon what you look for, what you seek to measure. In *Time's Arrow and Archimedes' Point,* Huw Price words it this way:

"Quantum mechanics tells us how the state function of a system changes over time... left to its own devices, its state function varies continuously and deterministically, in accordance with a mathematical rule known as Schrödinger's Equation. The only exception to this principle concerns what happens at measurement. At this stage, at least according to the orthodox interpretation of the theory, the state function undergoes a sudden, discontinuous, and generally indeterministic change, the nature of which depends in part on the nature of the measurement in question. If we measure the position of a particle, for example, its new state reflects the fact that what was measured was position and not some other property.

"...Much debate about the interpretation of quantum mechanics has turned on the significance of these two modes of evolution. Physicists inclined to the view that quantum mechanics provides a complete description of reality have thought of them something like this: before we measure, say, the position of a particle, its position is normally *objectively* indeterminate. It is not simply that the particle has a position which we don't know—if that were so the state function wouldn't be telling us the complete story—but that it has no determinate position at all. When we perform a position measurement on the particle, however, its state function "collapses" in such a way as to ensure that its position becomes determinate—it acquires a definite position, which it didn't have before."

Quantum reality, in other words, simply does not agree with human reality, human logic or common sense.

In David Lindley's book, *Where Does the Weirdness Go?,* he explains:

"The trouble with quantum mechanics, and the reason why there are "interpretations," is that despite decades of practical success with the theory, physicists still cannot honestly say that they know what the theory means; they can't see inside it, so to speak.  In the old days of classical physics, the formulas and equations clearly referred to an independent, objective world.  Particles existed, and had positions and speeds and energies that were well defined and unambiguous.  You might not know what they were, but you could be confident they were there.  Quantum mechanics, on the other hand, doesn't allow this.  Nothing is anything until you measure it; only then does it become a reliable, dependable something.  And worse, if you try to infer, from a given set of experimental data, what is going on beneath the surface—what is "really" there, underlying the measurements—you run the risk of finding yourself contradicted by someone else's view of what is really going on, someone who has used the same sort of data and the same kind of logic to infer, quite legitimately, a different "reality."  And so you conclude that reality depends on who is looking at it, or that there is no true reality at all.  Either conclusion seems to upset all the traditional notions of what scientists are looking for.

In short, physicists can use quantum mechanics with ease and assurance and yet not feel that they understand what is going on."

In our ordinary world, a particle is located at a specific point in space. A wave, on the other hand, is a disturbance spread out over a region of space. In the world of the very small, a quantum entity behaves as if it is a wave, if it is tested for wave-like properties, and it behaves as if it were a point particle, if it is tested for particle-like properties. Nobody really understands how this is possible. Further, as one might begin to suspect, you cannot measure for wave-like properties and particle-like properties *at the same time.*

In our world, a thing is either in this state, or in that state. It can be here, for example, or there; it can be in a stable state, or a state of decay; it can be twirling this way, or the opposite way. But in the quantum world, everything exists in a mixture of states: a particle is both here *and* there, it is stable *and* in a state of decay, it is twirling this way *and* it is twirling the opposite way. Only when a measurement is made does the quantum entity 'decide' to be in one certain state—it is here and not there—but it is in this state only while the measurement is being made; it then resumes being in an indeterminate mixture of states.

Nobody can explain how this occurs, or what actually, really is going on down there. There are guesses, of course, a countless number of theories and conjectures, but there is no scientific method of choosing between them.  Because nobody really knows: nobody really understands quantum mechanics.

The most common approach to this problem is the practical one: the equations of quantum mechanics work, and work extraordinarily well, so don't worry about what is 'really' going on down there—just do the equations and accept the results. As Feynman put it: "Do not keep asking yourself, if you can possibly avoid it, 'But how can it be like that?' because you will go 'down the drain' into a blind alley from which nobody has yet escaped. Nobody knows how it can be like that."

But physicists are people, and people generally want reality to make sense.  In *The Trouble with Physics*, Lee Smolin summarizes, "Quantum mechanics has been extremely successful at explaining a vast realm of phenomena. Its domain extends from radiation to the properties of transistors and from elementary-particle physics to the action of enzymes and other large molecules that are the building blocks of life.  Its predictions have been borne out again and again over the course of the last century.  But some physicists have always had misgivings about it, because the reality it describes is so bizarre.  Quantum theory contains within it some apparent conceptual paradoxes that...remain unresolved. An electron appears to be both a wave and a particle. So does light. Moreover, the theory gives only statistical predictions of subatomic behavior. Our ability to do any better than that is limited by the *uncertainty principle,* which tells us that we cannot measure a particle's position and momentum at the same time. The theory yields only probabilities. A particle— an atomic electron, say—can be anywhere until we measure it; our observation in some sense determines its state. All of this suggests that quantum theory does not tell the whole story. As a result, in spite of its success, there are many experts who are convinced that quantum theory hides something essential about nature that we need to know."

This sounds similar to the complaints we heard about relativity's frozen spacetime versus our perception of a flowing time. In fact, let us continue the comparison we began earlier between relativity and quantum mechanics.

In *The Elegant Universe*, Brian Greene tells us, "There are two foundational pillars upon which modern physics rests. One is Albert Einstein's general relativity, which provides a theoretical framework for understanding the universe on the largest of scales: stars, galaxies, clusters of galaxies, and beyond to the immense expanse of the universe itself. The other is quantum mechanics, which provides a theoretical framework for understanding the universe on the smallest of scales: molecules, atoms, and all the way down to subatomic particles like electrons and quarks. Through years of research, physicists have experimentally confirmed to almost unimaginable accuracy virtually all predictions made by each of these theories. But these same theoretical tools inexorably lead to another disturbing conclusion: As they are currently formulated, general relativity and quantum mechanics *cannot both be right*. The two theories underlying the tremendous progress of physics during the last hundred years—progress that has explained the expansion of the heavens and the fundamental structure of matter—are mutually incompatible."

These two theories have completely different physical and mathematical structures, so that the equations of one theory will not work in the other theory. And, perhaps more importantly, each theory makes fundamentally different assumptions about reality.

General relativity, for example, assumes physical reality is continuous. Quantum mechanics assumes physical reality is discrete. It is true that quantum mechanics has merged with *special* relativity into what is called a relativistic quantum field theory, but even here the discreteness of particles and energy remains.

General relativity describes reality as deterministic, which means the past determines the present and the future; quantum mechanics describes reality as probabilistic, which means the past only determines probable outcomes for the present and the future—and, again, the present and future already exist within relativity's spacetime, so there should be no probabilities involved.

In relativity theory, nothing can travel faster than the speed of light; this is one of its foundational pillars. But in quantum mechanics, influences travel instantaneously—and instantaneous travel unavoidably calls into question the very meaning of space and time.

For all practical purposes, then, general relativity and quantum mechanics seem to be describing two completely different realities. We are not referring, here, to the fact that general relativity describes large things, while quantum mechanics describes small things; no, large things are ultimately made up of small things, and small things ultimately make up large things; large and small things share a single reality. What we have here is a more fundamental problem than this. General relativity and quantum mechanics are so at odds with one another, *they are essentially describing two completely different realities.*

This presents us with an interesting

question.

What if they are?

What if general relativity and quantum mechanics *really are describing two completely different realities?*

What if, rather than assuming this is a problem to be solved, we assumed this was the solution to our problem?

In asking this question, in this manner, we have just grabbed a hold of spacetime's full head of hair.

So, let us start dragging.

Gently, at first...

# CHAPTOR FOUR

## Gently Dragging Spacetime

In relativity's spacetime, the past, present and future exist equally. Spacetime is a physical structure—a physical structure that is eternally unchanging. Space and time simply exist, as a fourth-dimensional spacetime continuum.

Time does not flow, within spacetime; there is no flowing river of time. Rather, spacetime is a frozen river, or a block of ice—it is a block universe.

We humans exist within the fourth dimensional spacetime continuum, as part of its structure. As such, we can precisely define our physical existence, within the spacetime continuum—we can define our existence as fourth-dimensional beings.

In three-dimensional space, we might define our physical human body as existing between, say, the top of our head and the bottom of our feet. In fourth-dimensional spacetime we likewise define

41

the "length" of our physical, human bodies—our fourth-dimensional beings—as extending between two "events." The first event we define as the moment of our conception, with the fertilization of a single cell; the last event we define as the moment of our "death," when our awareness ceases to be. Our fourth-dimensional beings exist between these two events—this is where we experience the world, live our life and are aware of existence. Throughout this life, we live believing we are mortal, and death is inevitable; however, within spacetime, our beings are complete, from beginning to end, simply and eternally existing.

We fear dying. We fear death. The thought of our awareness coming to an end, once and for all, permanently and forever, is a terrifying thought. But for over a hundred years, now, the scientifically formulated structure of relativity's spacetime has been telling us that our existence is eternal. We exist within spacetime, and we will always exist there. We will always be aware. There may come a time in the future, of course, in which we are not aware, but is this really different from a time in the past, before we were born, in which we were not aware? We are aware where we are aware, in time and in space, and that awareness is eternal. It will always exist.

Of course, this fact, by itself, does nothing

towards helping us solve the problems before us. It does not give us free will, or the ability to change our selves or our world, hopefully for the better.

In fact, what if our life is unpleasant, or actually unbearable? What if we know only hunger and disease, pain and suffering? Our life then would not be any type of eternal heaven, but a true living hell.

We have to ask, why do we even have the sensation of a flowing time, if it does not exist, and cannot exist? Why do we believe our awareness is moving forward through time, if it never has? Why do we believe that our actions affect the future, that our choices have consequences, and that we are responsible for those consequences? Why do we have such a strong belief that the future is unknown, and unpredictable?

Within spacetime, as we currently conceive it, all of these beliefs are simply wrong. They are illusions.

The puzzle is why these illusions exist.

A further, deeper puzzle is how awareness can exist at an instantaneous moment of time; how awareness can even have meaning without time's flow.

Our book seeks a solution to these puzzles. As we have warned from the beginning, we will do whatever it takes to find one.

Our basic premise, and our driving

assumption, is this: Our human perception of a flowing time is not an illusion. It is an accurate perception of physical reality.

Our goal is to have relativity's spacetime agree with human reality. To do this, we may have to prod spacetime, we may have to plead with it, we may have to re-visualize it, enlarge it, redefine it, or simply drag it cursing, kicking and screaming into a brand new conceptual reality.

One way or other, we are going to melt the frozen river of spacetime.

So, let us begin.

Gently, at first. Subtly.

Relativity unifies our three dimensions of space with our one dimension of time, producing the fourth-dimensional spacetime continuum. This description is usually accepted "as is," but let us analyze what this actually means.

Space has three dimensions. Essentially, this means that space has three *directions* that one can travel in: back and forth, left and right, and up and down. If we measure these dimensions, we might call them length, width and depth. A common question one encounters in popular science books is, why three? Why does space have three dimensions, rather than, say, two, or seven, or twenty-eight? A common answer that is given, or suggested (no one really knows the answer), is that if space had more or fewer dimensions, then life as we know it would not have evolved, and so there would not be any intelligent life forms around (to ask such intelligent questions).

For example, it has been shown that if space had, say, four dimensions, then planets would not form stable orbits around stars. There is, in fact, a whole list of difficulties that physical reality would have, if space had dimensions other than three. And so, this is probably why *our* space has three dimensions.

No effort is made to differentiate between these three spatial dimensions. We do not define length as the *first* dimension, for example, which might then put width as the second dimension, leaving depth as the third dimension. We do not do this because such definitions would be arbitrary. My length might be your width. People on other sides of the world—or on the moon—would define all these directions differently. So, the three spatial dimensions are simply assumed to be equivalent, but at right angles to one another.

Now let us turn our attention to *time*. Relativity unifies our three dimensions of space with our one dimension of time. But what, exactly, *is* time?

No one really knows the answer to this question. In relativity, as elsewhere, time is simply accepted as a given—just as our three spatial dimensions are accepted as a given. Time is a different *kind* of dimension than the space dimensions. It is a mysteriously different, other kind of dimension.

In the same manner that the three spatial dimensions are defined to be at right angles to each other, the time dimension is defined to be at a right angle to three-dimensional space. (Acceleration and gravity complicates these angles, however.)

Now, we are going to entertain a thought experiment. We are going to imagine an Alternate Universe, one that is slightly different from ours. In this alternate universe, there are people like us, perhaps alternate versions of ourselves. These alternate people have their own version of science, and of physics, with alternate assumptions and beliefs.

In this alternate universe, they have explored an entirely different interpretation of physical reality. Specifically, they have pursued a completely different concept of *dimensions*, the dimensions that make up their physical universe.

For example, they begin their concept of physical reality with the existence of a zero-dimensional point. They call this zero-dimensional point the *zeroth dimension.*

In their mind's eye, they then let this zero-dimensional point expand out, in opposite directions, becoming a one-dimensional line. They define this one-dimensional line as the *first dimension.*

The first dimension contains (has room for) an infinite number of zero-dimensional points.

They now let this one-dimensional line expand out, in opposite directions, becoming a two-dimensional plane. They call this two-dimensional plane the *second dimension.*

The second dimension has room for an infinite number of one-dimensional lines.

They now let this two-dimensional plane expand out, at a right angle, in opposite directions, becoming three-dimensional space. They call this three-dimensional space the *third dimension.*

The third dimension has room for an infinite number of two-dimensional planes.

Now, in the physical universe they envision as "reality," these four dimensions—the dimensions from zero to three—all exist, equally, making up the known spatial universe. Within their universe, in other words, there exist zero-dimensional points, one-dimensional lines, two-dimensional planes and three-dimensional space.

Their assumption is that these different dimensions might manifest themselves in somewhat strange, twisted forms in the actual physical universe. Two-dimensional planes, for example, might easily become distorted by physical phenomena into membrane like structures.

# THE STUBBORNLY PERSISTENT

We might note, here, that in this alternate universe, the question of why space has three dimensions never occurs to anyone, because space is simply *defined* to be three-dimensional. Reality exists as a hierarchy of dimensions, each higher dimension being a natural enlargement of the dimension below it. Space is three-dimensional, a plane is two-dimensional, a line is one-dimensional, and a point is zero-dimensional. And higher dimensional structures will have their own unique definitions.

Even though three-dimensional space is viewed as a natural enlargement of a two-dimensional plane, once a three-dimensional perspective is adopted, it becomes apparent that there are an infinite number of ways to divide the third dimension into two-dimensional planes (depending, perhaps, on the motion of a one-dimensional line within a two-dimensional plane that is moving within the third dimension).

The point is that the physicists in our imaginary alternate universe realize that reality depends upon the perspective they wish to adopt: which dimension they wish to explore. The third dimension will have a different 'reality' than the second dimension, for example.

They do not stop there, however. In their mind's eye, they now let three-dimensional space expand out, at a right angle, in opposite directions, becoming *the fourth-dimensional spacetime continuum.*

They define this fourth-dimensional spacetime continuum as the *fourth dimension.* The fourth dimension, in this reality, is a natural enlargement of the third dimension. It contains four dimensions, like in our reality, but unlike our reality this fourth dimension has a clear relationship to the dimensions below it. It is a higher dimension, in the same sense that the third dimension is a higher dimension than the two-dimensional plane.

Now, they also have an alternative way of describing this reality, one they find equally useful. The first dimension they define as allowing for the *movement* of zero-dimensional points. The second dimension they define as allowing for the movement of one-dimensional lines. The third dimension they define as allowing for the movement of two-dimensional planes—and, by extension, of one-dimensional lines, and zero-dimensional points. And the fourth dimension they define as allowing for the movement of three-dimensional space.

In this reality, the scientific reality they have created to explain their universe, nature remains logically consistent as she becomes more complex. The physicists in our alternate universe know there are alternative ways of describing dimensions, mathematically and physically, and they have explored these other concepts extensively, but they have concluded that *physical reality* actually manifests dimensions as a hierarchy, one natural enlargement at a time.

We might note, here, that in this alternate universe, there is nothing mysterious about time, or the fourth dimension. The fourth dimension is simply the next higher dimension, the one that allows for the movement of three-dimensional space. This happens to agree quite well with our human intuition of time, in which we equate movement with the flow of time—such as how long it takes to get somewhere, or the periodic oscillation of a clock—be it a grandfather clock, or a cesium atom. Even when we are standing still, we know the cells of our body are moving, and aging. If there were absolutely no motion—if all atoms were absolutely motionless—then there would be no passage of time. Both space *and* time would be frozen still. So, in the physics of our alternate universe, the fourth dimension allows for the movement of three-dimensional space, and, just like in our reality, the fourth dimension is *time.* Moving on to a fourth-dimensional perspective, however, the fourth dimension is seen as a succession of three-dimensional spaces, so that it can potentially contain an infinite number of three-dimensional spaces. From this fourth-dimensional perspective, then, these three-dimensional spaces become arbitrary, depending upon one's motion.

# GENTLY DRAGGING SPACETIME

We do not want to become too bogged down, here, in the physics of our imaginary alternate universe, so we will only mention just one of the technical details of this reality, in order to get a feel for their physics. Each enlargement of a lower dimension into a higher dimension involves the addition of a single new freedom of movement, at a right angle to the original dimension. For example, an infinite number of two-dimensional planes can pass through a one-dimensional line, so that the one-dimensional line could, in principle, travel in any direction. But the physicists in our alternate universe assume that only one particular, but arbitrary, two-dimensional plane gives a one-dimensional line its freedom of movement. The same is true for the enlargement of three-dimensional space—only one particular direction *through* the fourth dimension results in their fourth-dimensional spacetime continuum. These fascinating sidebars into their physics will only distract us from our goal, however, so we will remain focused on our target.

Our target, remember, is to somehow get *our* fourth-dimensional spacetime continuum to embrace a flowing river of time, and free will. At the moment, we are exploring the physics of an imaginary alternate universe, in order to see if such an existence is even possible, as a simple thought experiment. So far, all we have engaged in is a mere rewording of the situation, a shuffling of fundamental concepts, in order to see reality from a new perspective.

We have arrived at the point where the physical dimensions of our imaginary physics have enlarged up to the fourth-dimensional spacetime continuum. Our alternate physicists view their fourth-dimensional spacetime continuum as a natural enlargement or expansion of three-dimensional space, at a right angle, into the fourth dimension. They can either view this from a three-dimensional perspective, in which three-dimensional space is moving fourth-dimensionally, or they can adopt a fourth-dimensional perspective, in which the fourth-dimensional spacetime continuum consists of a succession of three-dimensional spaces (and those three-dimensional spaces can be arbitrarily divided into simultaneous hyperplanes).

*They do not stop there, however.*

They now let the entire fourth-dimensional spacetime continuum expand out, at a right angle to the fourth dimension, in opposite directions, becoming *the fifth-dimensional cosmic universe.*

This fifth dimension allows for the movement of the fourth-dimensional spacetime continuum. It is a natural enlargement of the fourth dimension.

Within our imaginary alternate universe,

they do not view their fourth-dimensional spacetime continuum as static and eternally unchanging. They do not do this because then they would have to ask strange questions of reality, such as, where is the flow of time in this universe? How can free will exist in a reality where nothing ever changes? In fact, why does this universe even exist? How could it simply, eternally exist, without any cause, without any origin?

No, they simply assume that *their entire fourth-dimensional spacetime continuum evolved up to its present state, and that it is continuing to evolve into a future state.*

Their fourth-dimensional spacetime continuum is evolving fifth-dimensionally.

This is a completely different concept of the fifth dimension than the one our physicists entertain. Before we continue the tale of this strange, imaginary universe, then, with its bizarre form of cosmology and physics, we will now return to our universe, our cosmology and physics, and find out how our physicists have grappled with the concept of the fifth dimension, specifically, and other dimensions in general.

# CHAPTOR FIVE

# Our Fifth Dimension

We will now briefly review the history of the fifth dimension, and other physical dimensions, as they evolved through various scientific theories, here in our universe. We will begin shortly after Einstein completed his general theory of relativity.

In Michio Kaku's book, *Hyperspace*, we find out that, "In April 1919, Einstein received a letter that left him speechless.

"It was from an unknown mathematician, Theodr Kaluza... In a short article, only a few pages long, this obscure mathematician was proposing a solution to one of the greatest problems of the century. In just a few lines, Kaluza was uniting Einstein's theory of gravity with Maxwell's theory of light by introducing the *fifth* dimension (that is, four dimensions of space and one dimension of time).

Now, if you are not paying close attention, this might easily become confusing. So take note.

Einstein united our three dimensions of space with our one dimension of time, which resulted in our fourth-dimensional spacetime continuum. (We are back in our real world, remember.) Now, along comes Kaluza, who postulates a fourth *spatial* dimension, giving space four dimensions, rather than three—plus, as before, the one dimension of time. Einstein's spacetime, however, had already defined four dimensions, so in order to talk about *another* dimension, through default this spatial dimension became the *fifth* dimension, even though it was actually a dimension of space (like the other three dimensions of space).

We would like to suggest, here, this was the beginning of a confusing terminology that has reverberated down to this day. (It may even, in fact, have been the point in which our reality diverged from our imaginary Alternate Universe.)

Putting this to the side, for now, let us continue with Michio Kaku's description of events:

"In this short note, Kaluza began, innocently enough, by writing down Einstein's field equations for gravity in five dimensions, not the usual four. (Riemann's metric tensor, we recall, can be formulated in any number of dimensions.) Then he preceded to show that these five-dimensional equations contained within them Einstein's earlier four-dimensional theory (which was to be expected) with an additional piece. But what shocked Einstein was that this additional piece was precisely Maxwell's theory of light. In other words, this unknown scientist was proposing to combine, in one stroke, the two greatest field theories known to science, Maxwell's and Einstein's, by mixing them in the fifth dimension."

There was an obvious problem to this idea—space only has three dimensions. If there were a fourth spatial dimension, someone would have noticed. *Everyone* would have noticed.

Kaluza was aware of this, of course, and proposed that the reason no one noticed this fourth spatial dimension was because it was in the shape of a small circle—a *very* small circle. This circle was smaller, in fact, than a grain of sand, it was smaller than the molecules making up the sand, it was smaller than the atoms making up the molecules—in fact, it was smaller than the nucleus inside of an atom. It was so small, then, it was invisible, for all practical purposes.

Our normal three dimensions of space, in contrast, are infinite in length, or at least practically infinite.

Kaluza gave no specific reason why this extra spatial dimension was so small, and in 1926 a mathematician, Oskar Klein, proposed that it had something to do with the newly formulated theory of quantum mechanics—that this new dimension was, in fact, the size of a Plank length—which was *really* small.

However, Michio Kaku ultimately concludes: "As promising as Kaluza-Klein theory was for giving a purely geometric foundation to the forces of nature, by the 1930s the theory was dead."

Neither Einstein, nor anyone else, could actually get the theory to work.

But, many years later, physicists became interested in this idea once again.

We will now let Lisa Randall pick up the story. In her book *Warped Passages*, she continues the tale of this small curled up dimension:

"So far, we've only considered a single additional dimension, which is rolled up into a circle. But everything we've said would hold true even if that curled-up dimension took some other shape— any shape at all. And it would also be true if there were two or more tiny, rolled-up dimensions of any shape at all. Any and all dimensions that are sufficiently small would be completely invisible to us."

Theories postulating the existence of very small, curled-up spatial dimensions have become known as Kaluza-Klein theories. They have gone through various incarnations, emerging in such theories as supergravity, supermembranes and superstrings.

Quantum mechanics began with the assumption that quantum particles are point particles, without any spatial extension. (If you find this concept puzzling, just be aware that you are not alone—how does something without any spatial extension even exist? But scientists have to assume *something* in order to proceed, and this assumption seems to work, and work well, despite it being conceptually contradictory.)

Then string theory came along, and it assumed point particles were really one-dimensional strings. (Actually, it assumed the vibration of the strings produced what we see as particles, but we are attempting to simplify things somewhat, here. Further, string theory merged with supersymmetry and became *superstring theory*, but it is commonly still referred to simply as string theory.)

Strings needed extra spatial dimensions to live in, for complicated mathematical reasons, and these spatial dimensions necessarily took complex shapes.

Lisa Randall continues with the evolving tale of these ideas:

"With more dimensions there are a huge number of conceivable *compact spaces*—spaces with rolled-up dimensions, distinguished by the precise manner in which the dimensions are rolled up. One category of compact spaces important to string theory are the *Calabi-Yau manifolds*... These geometric shapes roll up and wind together extra dimensions in a very special way. The dimensions are curled up into a small size, as with all compactifications, but they are tangled in a way that is more complicated and difficult to draw.

"Whatever shape the rolled-up extra dimensions take, and however many there are, at each point along the infinite dimensions there would be a small compact space containing all the curled-up dimensions. So, for example, if string theorists are right, everywhere in visible space...there would be a six-dimensional Calabi-Yau manifold of invisibly tiny size. The higher-dimensional geometry would be present at every point in space."

The concept of small curled-up dimensions did not stop there. Superstring theory (which was originally supposed to be The Theory of Everything) evolved into five *different* superstring theories, all describing a different ten-dimensional universe (four dimensions of spacetime, and six small spatial dimensions). Then superstring theory became M-theory, which has eleven dimensions—which was the same number of dimensions a once popular theory of supergravity had stumbled over, because of insurmountable anomalies. But M-theory was more robust, as Michio Kaku describes in his book, *Parallel Worlds*:

"M-theory can explain the existence of supergravity if we assume that a tiny portion of M-theory (just the massless particles) is the old supergravity theory. In other words, supergravity theory is a tiny subset of M-theory. Similarly, if we take this mysterious eleven-dimensional membranelike theory and curl up one dimension, the membrane turns into a string. In fact, it turns into precisely type II string theory! For example, if we look at a sphere in eleven dimensions and then curl up one dimension, the sphere collapses, and its equator becomes a closed string. We see that string theory can be viewed as a slice of a membrane in eleven dimensions if we curl up the eleventh dimension into a small circle."

Michio Kaku describes the continuing expansion of the concept of strings:

"One of the novel features of M-theory is that it introduces not only strings but a whole menagerie of membranes of different dimensions. In this picture, point particles are called "zero-branes," because they are infinitely small and have no dimension. A string is then a "one-brane," because it is a one-dimensional object defined by its length. A membrane is a "two-brane," like the surface of a basketball, defined by length and width. (A basketball can float in three dimensions, but its surface is only two-dimensional.) Our universe might be some kind of "three-brane," a three-dimensional object that has length, width, and breadth."

Kaku concludes:

"It can be shown that there are five ways in which to reduce eleven-dimensional M-theory down to ten dimensions, thereby yielding the five superstring theories."

In case you missed it, the branes of M-theory have an obvious resemblance to the hierarchy theory of dimensions that the physicists in our alternate universe accept as part of their physical reality. Except that, in our alternate universe, they have no interest in small curled-up spatial dimensions, and so they have no string theory.

As an interesting excursion, then, let us find out how *our* physicists view the concept of the lower dimensions—the dimensions from zero to three.

Surprisingly, perhaps, within our physics (apart from string theory), the lower dimensions are not even recognized as such, and so there has never been any attempt to organize them into a logical hierarchy. And yet clues to their existence, within our physical reality, are ubiquitous.

For example, we begin with the zero-dimensional point. In Robert Oerter's book, *The Theory of Almost Everything*, he tells us,

"It is tempting to think of electrons as tiny spinning balls of charge. There are problems with this picture, however. Think of how an ice skater speeds up as she pulls her arms in. The smaller she makes herself, the faster she spins. Now, no one has measured the size of an electron, but there are experiments that give it an upper limit. According to these experiments, the electron is so small that, to have the known value of spin, the surface of the "ball" would have to be moving faster than the speed of light. This, of course, is impossible. We are forced to conclude that an electron is not a tiny, spinning ball of charge...

"Well, then, what is an electron? What has charge and spin but isn't a spinning charge? The only option is to picture the electron as a truly fundamental particle, a pure geometric point having no size and no shape. We simply have to give up the idea that we can model an electron's structure at all. How can something with no size have mass? How can something with no structure have spin? At the moment, these questions have no answers. There is nothing to do but accept that the electron does have these properties. They are merely part of what an electron *is*. Until we find some experimental evidence of electron structure, there's nothing else to say about an electron."

Continuing up the hierarchy of physical dimensions, we encounter the one-dimensional line. Now, one might suppose that a ray of light is a good candidate for the manifestation of the first dimension, within physical reality, and indeed in our imaginary alternate universe this is the path they chose (we will describe their reasoning later in the book). But here in our universe they chose to investigate super tiny, one-dimensional strings.

And so, because strings are so popular, here, *and* because they are such a complicated subject, it is worth hearing a little more about these little critters, along with a brief summary of their complex history. We are not out to bedazzle you, here; we are merely trying to give you a feel for the issues involved. If you do not understand most of this, that is fine; you are in the company of almost everyone else on the planet. But take heart: if you can make it through the rest of this chapter, we assure you the remainder of the book will be a breeze, in comparison.

Like supergravity, string theory originally had troublesome anomalies—but these were eventually resolved, as Ian Stewart describes in his book, *Why Beauty Is Truth*:

"Green and Schwarz had discovered that very occasionally, the anomalies miraculously disappear, but only if space-time has 26 dimensions (in the first version of the theory, called bosonic string theory) or 10 dimensions (in later modifications). Why? In their calculations for bosonic string theory, the mathematical terms that would create an anomaly are multiplied by $d - 26$, where $d$ is the dimension of space-time. So these terms vanish precisely when $d = 26$. Similarly, in the modified version, the factor becomes $d - 10$. Time always remains one-dimensional, but space somehow acquires an extra 6 or 22 dimensions...

"If you were prepared to ignore the weird numbers 10 or 26, this discovery was very exciting. It suggested that there might be a mathematical reason for space-time to have a particular number of dimensions. It was disappointing that the number was not four, but it was a start. Physicists had always wondered why space-time has the dimensions it does; now it looked as though there might be a better answer to that question than, "well, it could be anything, but in our universe it's four"...

"M-theory posits an 11-dimensional spacetime, which unifies all five of the 10-dimensional string theories, in the sense that each can be obtained from M-theory by fixing some of its constants to particular values. In M-theory, Calabi-Yau manifolds are replaced by 7-dimensional spaces..."

Extra spatial dimensions, such as Calabi-Yau manifolds, are needed for strings to move and vibrate in—but what about the strings themselves? Roger Penrose explores the problem of point particles—how do dimensionless points "meet," in order to interact?—and how one-dimensional strings might solve some of these problems. In his book, *Fashion Faith and Fantasy in the New Physics of the Universe*, he elaborates:

"String theory suggests a different kind of answer to this conundrum. It proposes that the basic ingredient of matter is neither 0-dimensional in spatial extent, like a point particle, nor 3-dimensional, like a smeared-out distribution, but 1-dimensional, like a curved line. Although this may seem like a strange idea, we should bear in mind that, from the 4-dimensional perspective of spacetime, even a point particle is not described, classically, as simply a point since it is a (spatial) point which persists in time—so its space-time description is actually a *1-dimensional* manifold..., referred to as the *world-line* of the particle... Accordingly, the way that we should think of the curved line of string theory is as a 2-manifold, or *surface*, in space-time..., referred to as the string *worldsheet.*"

But Penrose is not happy with how string theorists ultimately fail to relate their theory with our actual physical reality:

"For example... If we are trying to think of these strings in a direct physical way...then we must think of a string's world-sheet as being *timelike*... If the world-sheet is to be without holes, then it must continue to be a timelike tube extending indefinitely into the future. It is no good thinking of it as wrapping around the "tiny" extra dimensions, since these dimensions are all taken to be spacelike. It can only continue indefinitely into the future, and then it does not really qualify as closed. This is one of many questions that I do not find to be properly addressed in any description of string theory that I have seen."

Roger Penrose makes the interesting point that a zero-dimensional particle is actually a one-dimensional world line, within spacetime, and a one-dimensional string becomes a two-dimensional manifold. This has an obvious (and relevant) resemblance to the hierarchy theory of dimensions, within our imaginary physics, and suggests how these concepts may have evolved—but we must leave that discussion for another time. But, later in our book, we will revisit the concept of lower dimensional objects extending indefinitely into the future.

Within the physics of our universe, mean-while, one-dimensional strings automatically lead into a discussion of two-dimensional surfaces, within spacetime. So, let us continue with this progression into the second dimension.

In Lawrence Krauss's book, *Hiding In the Mirror*, he similarly tells us,

"It turns out that because a string is a one-dimensional object moving in time, its "world sheet"—that is, the region of space-time it maps out as it moves—is a two-dimensional surface. This is the case whether the string is moving in four dimensions, ten dimensions, or twenty-six dimensions. Adding new fields onto the world sheet, which is what happens when fermions and Yang-Mills fields are added to strings, therefore involves studying how fields behave on two-dimensional surfaces.

"Interestingly, this is an area of intense interest in condensed matter physics, which studies the bulk properties of real material, whether boiling water, superconductors, or magnets. When such materials undergo a change of phase—for example, water begins to boil, magnets become magnetized—then near the point of this change the properties of the material become particularly interesting and simple. The physics turns out in some cases to depend almost entirely on phenomena associated with two-dimensional surfaces, such as bubble walls form the boundary between different phases of boiling water. As a result, condensed matter physicists have become experts on studying such surfaces...

"In any case, studies of such condensed matter systems had classified essentially all two-dimensional field theories, and demonstrated that many of them had the properties that one guessed they might have if they instead described string world sheets obtained by compactifying from higher dimensional theories. That is the good news. But at the same time it suggested that perhaps one could consider string theories in four dimensions without ever worrying about their ten-dimensional roots. Indeed, are the ten dimensions necessary at all, or are the extra dimensions just mathematical artifacts? This is the central question that continues to haunt us."

Now let us move on to the third dimension. In our imaginary alternate universe, they define "space" as three-dimensional. One might suppose this leaves little to talk about, in either of our realities—space is space—but what happens to three-dimensional space within the context of the fourth-dimensional spacetime continuum? Does three-dimensional space "lose its identity" within this higher dimension?

Julian Barbour actually discusses this question in his book, *The End of Time*:

"We are all familiar with flat surfaces (two-dimensional planes) in three-dimensional space. Such planes have one dimension fewer than the space in which they are embedded, and are flat. *Hyperplanes* are to any four-dimensional space what planes are to space. In Newtonian physics, space at one instant of time is a three-dimensional hyperplane in four-dimensional Newtonian spacetime. It is a *simultaneity hyperplane*: all points in it are at the same time. Such hyperplanes also exist in Minkowski spacetime, but they no longer form a unique family. Each splitting of space-time into space and time gives a different sequence of them.

"Now, what is Minkowski space-time made of? The standard answer is events, the points of four-dimensional space-time. But there is an alternative possibility in which three-dimensional configurations of extended matter are identified as the building blocks of space-time. The point is that the three-dimensional hyperplanes of relative simultaneity are vitally important structural features of Minkowski space-time. It is an important truth that special relativity is about the existence of distinguished frames of reference. And an essential fact about them is that they are "painted" onto simultaneity hyperplanes. As a consequence, simultaneity hyperplanes...are the very basis of the theory. They are distinguished features. You cannot begin to talk about special relativity without first introducing them...

"Minkowski space-time is not some amorphous bulk in which there is no simultaneity structure at all. We can 'paint coordinate lines'— and an associated simultaneity structure—on space-time in many different ways. But the whole content of the theory would be lost if we could not do it one way or the other. There is no doubt about it— simultaneity hyperplanes exist out there in spacetime as distinguished features."

Later in his book, Barbour discusses attempts to quantize general relativity: "One important approach, called *canonical quantization,* is based on analysis of the dynamical structure of the classical theory. This is how general relativity came to be studied in detail as a dynamical theory nearly half a century after its creation as a geometrical space-time theory. The 'hidden dynamical core,' or deep structure, of the theory was revealed...if general relativity is to be cast into a dynamical form, then the 'thing that changes' is not, as people had instinctively assumed, the four-dimensional distances within space-time, but the distances within three-dimensional spaces nested in space-time. The dynamics of general relativity is about three-dimensional things: Riemannian spaces."

Let us now return to the subject of "branes," as a sort of summary of the similarities between our physicists' reality, and our imaginary physicists' reality (as a hierarchy of dimensions).

In *Farewell to Reality*, Jim Baggott tells us,

"A ten-dimensional D-brane can be thought to consist of three 'conventional' spatial dimensions which extend off to infinity, six dimensions curled up and tucked away in a Calabi-Yau space, and time. The model demands that, once fixed to the D-brane, all the material particles of the standard model are then constrained to move in the three-dimensional space that we ourselves experience."

Of course, if you have one ten-dimensional D-brane, the next logical step is to imagine another, perhaps many others, separated by an eleventh (spatial) dimension.

Jim Baggott summarizes his thoughts on the matter:

"I suspect that your reaction to braneworld scenarios such as these is really a matter of taste. Perhaps you're amazed by the possibility that there might be much more to our universe than meets the eye; that there might exist dimensions 'at right angles to reality' that we can't perceive but whose influence is manifested in the behaviour of those particles that we can observe. The revelation that there might be multidimensional branes, bulk and hidden dimensions—large, small or warped—might prompt more than one 'Oh wow!' moment.

"But it doesn't do it for me, I'm afraid...

"My problem is that branes and braneworld physics appear to be informed not by the practical necessities of empirical reality, but by imagination constrained only by the internal rules of an esoteric mathematics and an often rather vague connection with problems that theoretical physics beyond the standard model is supposed to be addressing. No amount of window-dressing can hide the simple fact that this is all *metaphysics*, not physics."

Adding to these continuing difficulties with string theory are the overwhelming mathematical complexities involved, as Peter Woit discusses in his book, *Not Even Wrong*:

"Although different physicists may have varying opinions about the beauty of superstring theory, there is little disagreement about the difficulty of the theory. The simplest version of string theory...doesn't appear to have a stable vacuum, so one thing one needs to do is to consider a supersymmetric version involving fermions: the superstring. The basic equations for the superstring are complicated and come in several different versions, with the consistency now requiring 10 instead of 26 dimensions. Compactifying six of the dimensions brings in the full complexity of the geometry of curved six-dimensional Calabi-Yau spaces, a very difficult and challenging part of modern algebraic geometry.

"While ten-dimensional superstring theory compactified on a Calabi-Yau space is an extremely complex and difficult subject to master, in modern superstring theory it is just the beginning. The hope is that there is an underlying nonperturbative M-theory, but only a bewildering collection of partial results about this theory exist. Its low-energy limit is supposed to be eleven-dimensional supergravity, a rather complicated subject in itself, especially when one considers the compactification problem, which now requires understanding curved seven-dimensional spaces. Somehow, M-theory is supposed to describe not just strings, but higher-dimensional objects called branes. Many different sorts of calculations involving branes have been performed, but a fundamental theory capable of consistently describing them in an eleven-dimensional space with seven dimensions compactified still does not exist."

One might wonder, in the meantime, whatever became of Einstein, the one who came up with a theory that actually describes our universe. Alas, after his triumphant success with general relativity, he tried coming up with an even bigger, better theory, as Burton Feldman describes in his book, *112 Mercer Street*:

"Having geometrized gravitation, he now sought an even more general geometry that, while fitting gravity, would include electromagnetism as well. Riemann's geometry worked beautifully for enormous bodies, but could not be applied to atomic phenomena...

"Einstein had two choices. He could keep the Riemannian framework, but expand the number of dimensions to five or more. Or he could keep the four dimensions, but find a substitute for Riemann's geometry. At one time or another, decade after decade, he pursued both of these possibilities. He explored four- and five-dimensional continuums, differential geometries, gauge transformations, absolute parallelism. He took apart his final field equations for general relativity, assigning the symmetric part to gravitation and the antisymmetric part to the electromagnetic field. He spent the years puzzling at chalk marks on the blackboard...

"But unified theory was not to become

# OUR FIFTH DIMENSION

another stroke of genius and insight. It was more like an aging engine fitfully turning over... In 1945, at age sixty-six, he published his final equations, but hardly with the overwhelming confidence he had expressed about general relativity. When queried by reporters, he said, "Come back and see me in twenty years." He revised the equations in 1949 and 1954... Einstein himself wondered how definitive even his final equations were. Perhaps, he said wryly, his critics were right, and the equations did not "correspond to nature"—the ultimate defeat."

We stubbornly persistent are now faced with a dilemma. Superstring theory, or M-theory (or, perhaps soon, the theory formerly known as M-theory), appears to be stuttering out into a morass of infinitely possible universes—where you can pick any reality you prefer. Yet superstring theorists are *highly* intelligent and dedicated physicists. Even Einstein ultimately failed in his attempt to go beyond his general theory of relativity—and he was *really* smart, and *really* dedicated. What chance, then, do us mere mortals—us stubbornly persistent—have of going *beyond* the fourth-dimensional spacetime continuum, into a new, even better reality?

Our only chance is to go in the opposite direction of everyone else. Instead of searching for increasingly complex and esoteric mathematics and physics, we are going to try a far simpler approach—we are, in other words, going to dumb things down quite a bit. Our assumption is going to be that Einstein's relativity is a true description of reality *exactly as it is*—almost. Without changing relativity in any fundamental way, we are instead going to nudge it, just a little, giving it the smallest tweak possible to achieve our goal.

We are not out to solve the riddles of the universe. We are not striving to unify general relativity with quantum mechanics, or discover the true nature of time, or explain the feebleness of gravity, or quantum weirdness. No, our basic premise is going to be that everything we need to know is *already out there*, and all we need do is organize what is presently known into some new, holistic reality.

This reality must have free will, which is the single goal of our stubbornly persistent book. By extension, the spacetime continuum must somehow embrace a flowing river of time within its borders.

Once we catch a glimpse of this reality, we then will secure our passport, book our flight, survey the landscape, settle in, unpack our bags— and live there.

# CHAPTER SIX

## Super (Cosmic) Positions

In our imaginary alternate universe, physicists believe their physical reality consists of a hierarchy of dimensions, with each lower dimension enlarging naturally into the next higher dimension, gaining an additional freedom of movement along the way. With this physics, they have created an entirely different cosmology than ours. Specifically, they believe their fourth-dimensional spacetime continuum has the freedom of movement—the freedom to change and evolve.

In this reality, the fourth-dimensional spacetime continuum moves through their *fifth* dimension, which is a natural enlargement of their fourth dimension. This fifth dimension is a right-angled extension of their fourth dimension, and has room for an infinite number of fourth-dimensional spacetimes. *This is the reason why they call it the fifth dimension.*

As we explore this cosmology, we must first ask ourselves the most primary, basic question of them all, the one that must have occurred to everyone, by now.

Are there really alternate universes?

The concept of alternate universes, strangely enough, is actually quite popular among physicists, here in our real universe. Perhaps even more strangely, in our imaginary alternate universe they have abandoned all speculation about alternate universes.

The concept of alternate universes emerges naturally from physicists' attempts (here in our real universe) to understand quantum reality. Let us review what is so strange about quantum mechanics, which allows such wild speculation.

When Richard Feynman was still young (he would go on to win the Nobel Prize), he reformulated quantum mechanics into a new form, one more to his liking, called the "path-integral" or "sum-over-histories" approach. In the introduction to Feynman's book, *Six Easy Pieces*, Paul Davies tells us,

"The idea is that the path of a particle through space is not generally well-defined in quantum mechanics. We can imagine a freely moving electron, say, not merely traveling in a straight line between A and B as common sense would suggest, but taking a variety of wiggly routes. Feynman invites us to imagine that somehow the electron explores all possible routes, and in the absence of an observation about which path is taken we must suppose that all these alternative paths somehow contribute to the reality. So when an electron arrives at a point in space—say a target screen—many different histories must be integrated together to create this one event."

A quantum particle, in Feynman's approach, takes all of these different paths *simultaneously*, so that it actually interferes *with itself* along the way (bumping into itself, mind you, which affects its trajectory).

Say what?

In his book, *Dark Cosmos*, Dan Hooper explains:

"…we can no longer think of an electron as a clearly defined object, at one place at one time. Rather, an electron—and every other quantum object—is more like a wavy cloud of existence, smeared out over space and time… You cannot generally determine the precise position, velocity, or other property of a quantum object before it is measured, because the quantum object exists as a sort of combination of different states with different locations, velocities, and other properties. Stranger still, upon measuring the position of such an object precisely, you automatically lose any knowledge of its velocity, and vice versa. This ignorance is not the result of our limited vantage point or technical know-how, but is built into the very nature of every quantum object in the Universe: it is impossible for a particle-wave to have simultaneously a precise position and a precise velocity.

"Within the framework of quantum physics, the act of observation itself takes on a new and important role. Until an observation is made, quantum particles simply do not possess the well-defined characteristics that we expect of billiard balls and other macroscopic objects. These bizarre conclusions left many scientists very uncomfortable. To some, physics had undergone a transition from the somewhat unexpected to the completely counterintuitive and seemingly ridiculous."

Weird, strange, bizarre, crazy—and ridiculous.

What kind of reality is this?

In *The Road to Reality*, Roger Penrose grapples with the conflict between the mathematics of quantum mechanics and the reality the mathematics describe:

"It is a common view among many of today's physicists that quantum mechanics provides us with *no* picture of 'reality' at all! The formalism of quantum mechanics, on this view, is to be taken as just that: a mathematical formalism. This formalism, as many quantum physicists would argue, tells us essentially nothing about an actual *quantum reality* of the world, but merely allows us to compute probabilities for alternate realities that might occur. Such quantum physicists' ontology— to the extent that they would be worried by matters of 'ontology' at all—would be the view (a): that there is simply no reality expressed in the quantum formalism. At the other extreme, there are many quantum physicists who take the (seemingly) diametrically opposite view (b): that the unitarily evolving quantum state completely describes actual reality, with the alarming implication that practically all quantum alternatives must always continue to coexist (in superposition)...

"The supporters of alternative (b)...argue that when a measurement takes place, all the alternative outcomes actually *coexist* in reality, in a grand quantum linear superposition of alternate universes."

Alternate universes—with alternate observers.

In *Shadows of the Mind*, Roger Penrose explores this concept further:

"In the many-worlds-type viewpoints, then, there would be different instances (copies) of the observer's 'self', co-existing within the total state and having different perceptions of the world around. The actual state of the world that accompanies each copy of the observer would be consistent with the perceptions of that copy.

"We can generalize this to the more 'realistic' physical situations in which there would be huge numbers of different quantum alternatives continually occurring throughout the universe's history... Thus, according to this many-worlds type of viewpoint, the total state of the universe would indeed comprise many different 'worlds' and there would be many different instances of any human observer. Each instance would perceive a world that is consistent with that observer's own perceptions, and it is argued that this is all that is needed for a satisfactory theory."

If an experiment is designed to find out if an electron is, say, spinning up, or spinning down, then, according to the many-worlds interpretation, one universe will record it as spinning up, and the other universe will record it as spinning down. There will be two universes, with different results (and two different copies of the observers within these universes). But there are many experiments that can be performed on an electron, with different possibilities, and a real electron is interacting with a real universe on a continuing basis, performing "experiments" of its own, with various possible outcomes.

Paul Davies discusses this situation in his book, *The Mind of God*:

## SUPER (COSMIC) POSITIONS

"In general, one would need an infinite number of universes to cover all possibilities.

"Now imagine extending this idea from a single electron to every quantum particle in the universe. Throughout the cosmos, the inherent uncertainties that confront each and every quantum particle are continually being resolved by differentiation of reality into ever more independently existing universes. This image implies that everything that can happen, will happen. That is, every set of circumstances that is physically possible (though not everything that is logically possible) will be manifested somewhere among this infinite set of universes.

"The various universes must be considered to be in some sense "parallel" or coexisting realities. Any given observer will, of course, see only one of them... It is part of the theory that you can't detect this mental "splitting"; each copy of us feels unique and integral. Nevertheless, there are stupendously many copies of ourselves in existence! Bizarre though the theory may seem, it is supported, in one version or another, by a large number of physicists as well as some philosophers. Its virtues are particularly compelling to those engaged in quantum cosmology, where alternative interpretations of quantum mechanics seem even less satisfactory. It must be said, however, that the theory is not without its critics, some of whom (e.g.,

Roger Penrose) challenge the claim that we would not notice the splitting."

In *Schrödinger's Kittens and the Search for Reality*, John Gribbin ponders these ideas further:

"The basic idea of the many-worlds theory is that every time the Universe is faced with a choice at the quantum level, the entire Universe splits into as many copies of itself as it takes to carry out every possible option...

"The problem is that in its original form the many-worlds interpretation requires an infinite number of universes, each splitting into infinitely more versions of reality every split second, as all the atoms and particles in the universe(s) are faced with quantum choices and follow every possible route into the future at once. The usual way to think of this splitting of universes is in human terms— that there might be a 'parallel world' in which the South won the American Civil War, one in which the communists never seized power in Russia, and so on. As I said, delightful stuff for the science fiction writers, and seemingly reasonable enough, at this human level... But is it so reasonable if we have to allow for every single tiny quantum choice to turn out in every possible way? And if that isn't reasonable, but the big choices affecting human history are the cause of a proliferation of universes, then we are back to the problem of deciding where you draw the line between the quantum world and the everyday world, and puzzling over whether or not the implications of a quantum choice have to be big enough for an intelligent observer to notice before it has any effect."

The central problem running throughout this discussion involves the role of an 'observer.' When an observation of a quantum particle occurs, the system changes. When unobserved, the system progresses according to specific probabilities; but when the system is actually observed—by some appropriate detection device, or by an actual person— only one of those probabilities is recorded (or sensed). So, either the observer is causing reality to "choose" a definite outcome, or else all outcomes remain equally real—and the observer has split into as many outcomes as the probabilities predict. This, essentially, defines the conflict between the original Copenhagen interpretation of quantum mechanics, and the many-worlds interpretation.

In his book, *The Cosmic Landscape*, Leonard Susskind asks,

"What if Germany had won World War II? Or what would life be like if the asteroid that killed the dinosaurs sixty-five million years ago had not hit the earth? The idea of a parallel world that took a different course at some critical historical junction is a favorite theme of science-fiction authors. But as real science, I have always dismissed such ideas as frivolous nonsense. But to my surprise I find myself talking and thinking about just such matters...

"I am far from the first physicist to seriously entertain the possibility that reality—whatever that means—contains, in addition to our own world of experience, alternate worlds with different history than our own. The subject has been part of an ongoing debate about the interpretation of quantum mechanics. Sometime in the middle 1950s, a young graduate student, Hugh Everett III, put forth a radical reinterpretation of quantum mechanics that he called the *many-worlds interpretation*. Everett's theory is that at every junction in history the world splits into parallel universes with alternate histories. Although it sounds like fringe speculation, some of the greatest modern physicists have been driven by the weirdness of quantum mechanics to embrace Everett's ideas...

"By the time we get to the present stage of history, the wave function has branched so many times that there are an enormous number of replicas of every possible eventuality... The number of branches containing you, sitting and reading this book, is practically infinite."

So, greetings, all of you out there.

At the risk of becoming repetitive—although all these quotes will have interest for us in the next chapter—let us hear from Paul Davies once more, this time from his book, *God & the New Physics*:

"Commonsense may rebel against the extraordinary concept of the universe branching into two as the result of the antics of a single electron, but the theory stands up well to closer scrutiny...

"But where *are* these worlds? In a sense, those that closely resemble our own are very nearby. Yet they are totally inaccessible: we cannot reach them however far we travel through our own space and time. The reader of this book is no more than an inch away from millions of his duplicates, but that inch is not measured through the space of our perceptions.

"The farther apart the worlds have branched, the greater their differences. Worlds that split away from our own in some trivial way, such as the path of a photon in a two-hole experiment, would be indistinguishable to the casual glance. Others would differ in their cat populations. In some worlds Hitler would not have been, John Kennedy lives on. Yet others would be wildly different, especially those that branched away from each other near the beginning of time. In fact, everything that could possibly happen (though not everything that can conceivably happen) does happen somewhere, in some branch of this multifoliate reality."

# THE STUBBORNLY PERSISTENT

We now offer one final, intriguing thought on the subject of alternate universes. This is from Paul Halpern's book, *Edge of the Universe*:

"Feynman's technique was one of the first mathematical expressions of the idea of parallel realities...

"Feynman did not intend his method to represent actual parallel realities; he was too practical for that. Rather, he meant it as mathematical shorthand for how quantum measurements must take into account the principle of uncertainty. The roads all taken represented the lack of knowledge within our world, rather than bifurcations into alternative worlds. However, another of Wheeler's students, Hugh Everett III, would make that leap...

"He conjectured that each time a subatomic event occurs—whether it represents decay, scattering, absorption, or emission—the universe bifurcates into parallel realities. Not only does the quantum interaction split into distinct realities, everything else does, too. Hence anyone observing a quantum experiment would witness a result that depends on exactly which version of truth he or she happens to be in...

"Everett...died suddenly in 1982 of a heart attack at age fifty-one...

"Until the end, because of his ardent faith in parallel universes, Everett had believed that death was impossible—a philosophy that has come to be known as quantum immortality. Each quantum process, he thought, would lead to a splitting-up of a person's conscious identity. Therefore even if one of the copies dies, others would live on. With further quantum transitions, the surviving copies would then bifurcate again and again in a never-ending progression. Whenever the candle blew out for any of the versions there would always be others left to carry on the flame of consciousness."

So, there we have it. Now that we have the concept of alternate universes firmly established (as a legitimate scientific speculation), let us return to our imaginary alternate universe, and find out how they are doing.

# CHAPTOR SEVEN

## Imaginary Times

In our real universe, physicists have argued extensively over the merits and demerits of the many-worlds interpretation of quantum mechanics. According to this view of reality, our universe is surrounded by an infinite number of alternate universes. Nearby universes are indistinguishable from our own universe; farther away universes show progressively more differences. And all of these universes trace their origin back to the Big Bang.

In a way, this scenario resembles our ordinary view of the passage of time. The universe of an infinitesimal moment ago, and the universe of an infinitesimal moment before that, and the universe of an infinitesimal moment from now, are practically indistinguishable from the universe of right now (and there are an infinite number of infinitesimal moments within any time interval), while the universe of yesterday, and tomorrow, will show definite changes from the universe of today. The farther away in time, the more changes there are.

Also, the universe of a moment ago is unreachable for us, in much the same manner that a universe that records an electron as spin up, while we have recorded it as spin down, is unreachable for us.

Perhaps, as David Deutsch suggests (at the beginning of our book), the resemblance between other times and other universes is more than a simple coincidence. Perhaps other times *are* other universes.

Or, perhaps, there is another possibility.

If we take the concept of alternate universes seriously—which we will do, for the moment, for the sake of our imaginary alternate universe—then the history of physics, itself, as well as the history of cosmology, will have taken many different paths in many different alternate universes. There will, in fact, be an infinite number of different paths, following every conceivable alternative.

We are going to take advantage of this fact, and assume that any cosmology we imagine—as long as it is logically consistent—will actually exist out there as the cosmology of some alternate universe.

# IMAGINARY TIMES

This does not imply that the cosmology is correct, of course—there are many examples in our own history where even strongly held beliefs turned out to be wrong. The concept of an ether, for example, was widely accepted as a confirmed fact at the beginning of the 1900s, and persisted in the minds of many distinguished scientists even after Einstein's special theory of relativity had done away with it. So just because we assume our cosmology exists out there, somewhere, is in no way meant to suggest it is *correct.* These alternate scientists may have gotten it all wrong.

What we are offering is simply a thought experiment, nothing more.

In our imaginary alternate universe, they believe their fourth-dimensional spacetime continuum is a natural, right-angled extension of three-dimensional space. In comparison to our physics, this turns out to be more of a conceptual difference than a practical one; physicists in our imaginary alternate universe firmly believe that Einstein's theories of relativity (the special and the general theory) accurately describe their fourth-dimensional spacetime continuum.

Specifically, they believe these things about physical reality:

*Any two points in spacetime can be thought of as being arbitrarily close together—as close together as one wishes to imagine.* Physical reality, in other words, is *continuous*. This is where the word *continuum* fits in—the fourth dimensional spacetime *continuum.*

In *Relativity—The Special and the General Theory*, Einstein words it like this:

"Space is a three-dimensional continuum. By this we mean that it is possible to describe the position of a point (at rest) by means of three numbers (co-ordinates) *x, y, z,* and that there is an indefinite number of points in the neighborhood of this one, the position of which can be described by coordinates...which may be as near as we choose to the respective values of the co-ordinates *x, y, z* of the first point. In virtue of the latter property we speak of a "continuum," and owing to the fact that there are three co-ordinates we speak of it as being "three-dimensional."

"Similarly, the world of physical phenomena which was briefly called "world" by Minkowski is naturally four-dimensional in the space-time sense. For it is composed of individual events, each of which is described by four numbers, namely, three space co-ordinates *x, y, z* and a time co-ordinate, the time-value *t.* The "world" is in this sense also a continuum; for to every event there are as many "neighbouring" events (realised or at least thinkable) as we care to choose, the co-ordinates...of which differ by an indefinitely small amount from those of the event *x, y, z, t* originally considered."

Spacetime is continuous.

Moving on to the next point, the physicists in our alternate universe believe:

*The past, the present and the future equally exist.*

Putting the above two points together means that: *All three-dimensional objects in the universe exist continuously through time*, meaning they are actually fourth-dimensional objects. They have an extension in time as well as in space. Fourth-dimensional objects have a real, physical existence—in fact, they are more 'real' than their three-dimensional 'snapshots' that we perceive.

Because the past, the present and the future have an equal existence, *the spacetime continuum exists as a single, complete, physical structure.* This structure was smaller in the past than it is today—it is continuously smaller, the farther back in time one travels, shrinking all the way down to an infinitesimal point at the beginning of time. In the other direction, the universe grows progressively larger into the future, expanding out, and out, and as it expands all of the matter and energy in the universe becomes ever more spread out and diluted, until, finally, far far into the future, the entire universe, for all practical purposes, simply ceases to exist. (Alternatively, the universe might eventually reverse its expansion, and ultimately collapse, but at the moment the first alternative is the popular view, so we will stick with this description for now. The second alternate would not change the following scenario in any fundamental way.)

The fourth-dimensional spacetime continuum, then, has a specific shape. *It is small at one end, and grows progressively larger towards the other end.* A simple analogy would be to a cone, perhaps; an ice cream cone that is small at one end and grows progressively larger towards the other end. Or one could compare it to a three-dimensional ball, which has a center, which can be defined as an infinitesimal point, and grows progressively larger, the farther out from the center one travels. But we need to keep in mind that these are only analogies, and that the spacetime continuum is actually fourth-dimensional—it has four dimensions, not three.

If all fourth-dimensional objects in the universe exist continuously through time, a natural question is, how far do they extend? We have already discussed our fourth-dimensional beings, whose fourth-dimensional bodies extend from the moment of our conception, as a single cell, to the moment of our "deaths." A little thought, however, reveals this to be an arbitrary definition, if we consider what actually makes up the human body.

In his book, *One Two Three...Infinity*, George Gamow discusses this concept. (His book is a classic in both realities, of course.) In the book,

# IMAGINARY TIMES

he describes the fourth-dimensional, physical existence of the human body, as it extends through spacetime, comparing it to a long rubber bar. Because it is impossible to draw fourth-dimensional objects on a two-dimensional page, he refers to a picture in the book of a two-dimensional "shadow man," whose existence extends through time:

"Think of yourself as a four-dimensional figure, a kind of long rubber bar extending in time from the moment of your birth to the end of your natural life... The picture represents just a small section of the entire life span of our shadow man. The entire life span should be represented by a much longer rubber bar, which is rather thin in the beginning, when the man is still a baby, runs wiggling through the period of many years of life, attains a constant shape at the moment of death (because the dead do not move), and then begins to disintegrate.

"To be more exact we must say that this four-dimensional bar is formed by a very numerous group of separate fibers, each one composed of separate atoms. Through the period of life most of these fibers stay together as a group; only a few of them fall away, as when the hair or the nails are cut. Since the atoms are indestructible, the disintegration of the human body after death should be actually considered as the dispersion of the separate filaments (except probably those forming the bones) in all different directions.

# THE STUBBORNLY PERSISTENT

"In the language of four-dimensional spacetime geometry the line representing the history of each individual material particle is known as its "world-line." Similarly we can speak of the "world-bands" composed of a group of world-lines forming a composite body."

This brings up an interesting point. The human body is made up of atoms, and it is continuously gaining and shedding different atoms throughout its existence. In fact, all three- and fourth-dimensional objects are made up of atoms, and it is the atoms that extend continuously through time. And atoms are indestructible (up to a point). So how far in time do these atoms extend?

Atoms exist continuously into the future, perhaps indefinitely into the future. Or, perhaps, after some indescribably long period of time, the protons and neutrons that make up the atoms, as well as the quarks making up the protons and neutrons, will simply dissolve into their equivalent measure of energy—which itself will continually dilute and fade away with the passage of eons. Although it would be fascinating to pursue this subject (for the interested, Roger Penrose explores this concept in his book, *Cycles of Time*), for our purposes we need to explore the other direction of time, to discover where atoms first originated.

Tracing the history of atoms continuously back through time, one ultimately ends up inside of stars, where most atoms are forged (except for the lightest ones). And stars themselves are also fourth-dimensional objects, which disperse their matter and energy out into the universe, ultimately reforming into planets and people. Meanwhile, inside of stars, larger atoms are made from smaller atoms, and from energy—and mass and energy are equivalent, as Einstein himself told us. So, we can follow all of these mass-energy structures back further and further in time, past the formation of stars, into the earliest stages of the universe.

In fact, we can continue tracing these mass-energy structures all the way back through time—back to the very beginning, to the Big Bang itself. It is here that all energy (and thus all mass) sprang into being. (At least, this marks the boundary of their physical existence; as we pointed out at the beginning of our book; the wording here is immaterial.)

This means that all fourth-dimensional objects in the universe stretch continuously back through time, changing shape and form, mass and energy, to the very beginning of the universe. Here, at the Big Bang, all fourth-dimensional objects ultimately meet, and merge together. This is where they all began—this is where they all originate.

The infinitesimal point at the beginning of time, known as a singularity, is where all of the fourth-dimensional objects in the universe unite as one. Now, maybe this singularity is where the laws of the universe break down, or maybe this point is somewhat smeared out in space and time, so that it is not actually a true singularity. We will discuss this alternative in a moment, but for now we focus on the fact that this is where all fourth-dimensional objects originated, which means that at this singular point they all meet and merge into a single spacetime point of energy.

Because they meet, merge and join together at this single point at the beginning of time, in a very real sense, then, all four-dimensional objects in the universe actually form a *single* fourth-dimensional object. The singularity at the beginning of time unites all fourth-dimensional objects into a single object, which then extends out through space and time.

All the matter and energy in the entire universe, in other words, forms a single, fourth-dimensional object that extends continuously through time.

*We do not stop there, however.*

Taking the concept of continuity *seriously* allows us to extend this concept to the next higher level of description.

Because all energy, and thus all mass, in fact all of spacetime itself, originated at the singular point at the beginning of time, and then extended *continuously* out, to the present day, and far into the future, this means that the entire fourth-dimensional spacetime continuum is not simply a single, fourth-dimensional object—it is a single, fourth-dimensional *particle*. It exists continuously through spacetime as a single, fourth-dimensional particle. The concept of continuity not only allows this description, but actually requires it. It is a simple extension of the concept of an ordinary, physical particle, such as an electron, which extends continuously through time, except that now we are referring to the entire fourth-dimensional spacetime continuum. An electron at different moments in time is the *same* electron, extending fourth-dimensionally through time. Simply enlarging this concept, the fourth-dimensional particle (that is the entire universe, made of electrons, etc.) is the same, single particle extending through time.

And this is how the physicists and cosmologists in our imaginary alternate universe view reality. They think of their fourth-dimensional spacetime continuum as a single fourth-dimensional particle. This particle consists of matter, energy, and a surrounding spacetime continuum. Only three-dimensional slices of this fourth-dimensional object are visible to us at any one moment of time. But we humans are also a part of the particle—we are embedded within its structure. The fact that observers arbitrarily divide this particle into different definitions of past, present and future reinforces the concept that it is the *same* particle being observed differently by different observers, depending upon their orientation within the particle.

We need to stress that we have introduced no new physical phenomena, here; all we have engaged in is a novel rewording of our presently known physical universe. Although it is a different way to view reality, it still accurately describes our familiar fourth-dimensional spacetime continuum. It simply provides a new perspective.

Now, in our universe, a certain Steven Hawking figured out a way to get rid of the singularity that general relativity predicts must exist at the beginning of time. In his book, *Cosmic Jackpot*, Paul Davies first describes this singularity, as one travels back to the Big Bang:

"The real universe consists of more than expanding space, of course: there is matter too. As space is compressed to zero volume, the density of matter becomes infinite, and this is so whether space is infinite or finite—in both cases there is infinite compression of matter to an infinite density. In Einstein's general theory of relativity, on which this entire discussion is based, the density of matter serves to determine (along with the pressure) the curvature or distortion of spacetime. If the theory of relativity is applied uncritically all the way down to the condition of infinite density, it predicts that the spacetime curvature should also become infinite there. Mathematicians call the infinite curvature limit of spacetime a *singularity*. In this picture, then, the big bang emerges from a singularity."

Paul Davies then describes how the "Hartle-Hawking" proposal eliminates the singularity. He refers to a picture in his book, which shows the universe shrinking down to the singularity:

"In this picture, spacetime looks like an upside-down cone, but apart from the sharp apex at the base, not too much importance should be placed on the shape. The effect of quantum uncertainty is to change the structure of the cone near the apex... The infinitely sharp point, representing a spacetime singularity, gets replaced by a rounded bowl. The radius of this bowl is about a Plank length, extraordinarily tiny by human standards, but not actually zero—which is the crucial thing. The singularity has therefore been removed in this description.

"Translated into spacetime language, and using our familiar play-the-movie-backward description, this diagram describes a universe that contracts inexorably toward a zero radius, which it is destined to reach at a certain predicted time, but just before this singular terminating event happens (about one Plank time before it), time itself starts to get fuzzy, seized by an identity crisis, and begins to adopt more and more spacelike qualities. Time doesn't change abruptly into space—the proposed Hartle-Hawking mathematical construction sees to that— but fades away in a continuous manner. At "the base of the bowl" time has become purely spacelike. Adopting forward-in-time terminology, this says that, at the start, there were actually four dimensions of space, one of which transformed itself into time. This transformation wasn't a sudden "switching on" of time...although in human terms it was rapid enough, occupying only about a Plank time (or rather it would have done, had time properly existed). But crucially, it was not instantaneous. A singular origin of the universe, the event without a cause that seemed to place the cosmic origin outside science, is replaced in this theory by a smooth origin, complying with the laws of physics everywhere."

Davies then goes on to say how the Hartle-Hawking proposal, while an important first step in new cosmological thinking (at the time), should not be taken too seriously. Indeed, over the years it has encountered various difficulties, but what interests us here are its conceptual basis'—particularly its use of the concept of *imaginary time.*

In *The Universe in a Nutshell*, Stephen Hawking explains:

"Because the universe keeps on rolling the dice to see what happens next, it doesn't have just a

# IMAGINARY TIMES

single history, as one might have thought. Instead, the universe must have every possible history…

"We are working to combine Einstein's general theory of relativity and Feynman's idea of multiple histories into a complete unified theory that will describe everything that happens in the universe. This unified theory will enable us to calculate how the universe will develop if we know how the histories started. But the unified theory will not in itself tell us how the universe began or what its initial state was. For that, we need what are called boundary conditions, rules that tell us what happens on the frontiers of the universe, the edges of space and time.

"If the frontier of the universe was just at a normal point of space and time, we could go past it and claim the territory beyond as part of the universe. On the other hand, if the boundary of the universe was at a jagged edge where space and time were scrunched up and the density was infinite, it would be very difficult to define meaningful boundary conditions.

"However, a colleague named Jim Hartle and I realized there was a third possibility. Maybe the universe has no boundary in space and time. At first sight, this seems to be in direct contradiction with the theorems that Penrose and I proved, which showed that the universe must have had a beginning, a boundary in time. However…there is another kind of time, called imaginary time, that is at right angles to the ordinary real time that we feel going by. The history of the universe in real time determines its history in imaginary time, and vice versa, but the two kinds of history can be very different. In particular, the universe need have no beginning or end in imaginary time. Imaginary time behaves just like another direction in space. Thus, the histories of the universe in imaginary time can be thought of as curved surfaces, like a ball, a plane, or a saddle shape, but with four dimensions instead of two."

As Hawking explains above, imaginary time exists at right angles to real time.

In his book, *The Void*, Frank Close gives us this view:

"We have already seen Einstein with his four-dimensional picture of space-time, with curvature related to gravity. Hawking and Hartle have gone further and imagine the universe as a four-dimensional surface of a five-dimensional sphere. I cannot visualize this, nor to be fair do its authors other than mathematically. However, we can visualize a simpler version, playing once again the role of creatures who are aware only of limited spatial dimensions whose universe is perceived to be expanding in time. This will suggest that our universe only appears to be expanding as a result of our limited cognition. In the Hawking-Hartle model, there is no expansion, no beginning: the universe simply exists...

"It is possible to imagine that what we call the Big Bang was when the compact universe emerged from the era of quantum gravity, which is when time took over from imaginary time. Questions about where everything came from, how it all 'began', are sidestepped; the universe in this picture has no beginning, no end: it just is."

In *Black Holes and Baby Universes*, Stephen Hawking elaborates further:

"Part of Einstein's problems with quantum mechanics and the uncertainty principle arose from the fact that he used the ordinary, commonsense notion that a system has a definite history. A particle is either in one place or in another...

"An elegant way to avoid these paradoxes that had so troubled Einstein was put forward by the American physicist Richard Feynman... The idea was that a system didn't have just a single history in space-time, as one would normally assume it did in a classical nonquantum theory. Rather, it had every possible history...

"You can think of ordinary, real time as a horizontal line, going from left to right. Early times are on the left, and late times are on the right. But you can also consider another direction of time, up and down the page. This is the so-called imaginary direction of time, at right angles to real time."

# THE STUBBORNLY PERSISTENT

In Hawking's vision, our universe is surrounded by an infinite number of alternate universe's that exist in imaginary time.

Now, in our imaginary alternate universe, there is another version of Stephen Hawking. (He lives on, much like the immortal Elvis does in our universe.) This alternate Stephen Hawking came up with an alternate version of his no-boundary proposal, one more in line with current cosmological thinking (the current thinking of the other cosmologists in his alternate universe). Instead of proposing that our (his) universe is only one of an infinite number of universes existing in imaginary time, this Stephen Hawking instead proposed that there is only one universe—his universe—and this universe is *moving through imaginary time.*

His entire fourth-dimensional spacetime continuum is moving. It is moving at a single, right-angled extension through imaginary time. (Note that this is in contrast to the above description, in which imaginary time exists at right *angles* to real time.)

Cosmologists in his universe call this right-angled extension the *fifth-dimensional cosmic universe.*

The reason Hawking came up with this version of his no-boundary proposal was it allowed him to solve several conceptual difficulties.

# IMAGINARY TIMES

For example, with this proposal Hawking realized he could eliminate the concept of alternate universes altogether. Instead of alternate universes, they were actually the *fifth-dimensional past histories of the evolving fourth-dimensional spacetime continuum.*

They were the past histories of the universe, as the universe moved and evolved through the fifth dimension.

With a wave of his hand, this alternate Stephen Hawking banished all of those infinitely branching universes, and made his own universe special, once again.

But this was only the beginning of his proposal. Alternate Hawking sought to answer certain questions, such as, What was the nature of this movement? How was the spacetime continuum moving? Where was it moving, why was it moving?

Contemplating all of the known facts about spacetime (many of which we have listed above), our alternate Stephen Hawking realized there was only one obvious, inevitable conclusion.

The entire fourth dimensional-spacetime continuum must be spinning about its center (seen as the beginning of time), through imaginary time.

The fourth-dimensional spacetime continuum was spinning fifth-dimensionally.

# THE STUBBORNLY PERSISTENT

Now, by this point it may seem as if our alternate reality has plunged into fantasy-land, which, of course, it has. But be assured: there will be no more wild and crazy ideas introduced, for the rest of the book. So the only true question we now face is, Is this concept crazy enough?

The remainder of our tale will be focused on organizing our weird reality into a single, holistic, comprehensible reality—a reality in which, hopefully, we have melted the frozen river of spacetime, and free will is an integral part of human reality.

# CHAPTER EIGHT

## Quantum Continuums

We will now explore the consequences of our alternate Stephen Hawking's cosmology, in which the entire fourth-dimensional spacetime continuum is spinning about its center, fifth-dimensionally, through imaginary time.

If the entire fourth-dimensional spacetime continuum is moving, this means that the entire fourth-dimensional spacetime continuum *has a past*. This is not the past one normally assumes, which is a fourth-dimensional past of three-dimensional space; no, this is a fifth-dimensional past of fourth-dimensional spacetime.

The fourth-dimensional spacetime continuum, in this cosmology, is moving. It is changing and evolving, fifth-dimensionally. It evolved up to the present, and it is continuing to evolve into a future state.

This solves the problem of how a static, eternally unchanging fourth-dimensional spacetime continuum ever got here in the first place.

It evolved here from somewhere else.

It had past states of existence, it has a present state of existence, and it will have future states of existence.

Those past states of existence are what we in our universe imagine to be alternate universes. The people in our alternate universe see them as past universes.

Admittedly, this concept potentially opens up an existential crisis for normal humans, in either universe. If the universe is changing and evolving—all of its past, present and future—then what becomes of *our* past, present and future?

Let us hear how the physicists in our alternate universe describe the situation:

Fortunately, our present remains effectively unaltered within this cosmology, because this is how we view the present anyway—as changing and evolving, open and free. In fact, the same can be said for how we view the future—as open, free and objectively indeterminate. The future might 'already' exist, but that future is changing as much as the present—and, that future is largely dependent upon what happens in the present, as these influences travel out at the speed of light (at the most).

This means that we have aligned these two modes of time (the present and the future) so that they are now in agreement with our own human perception of reality.

(Wait—wasn't that the goal of our book—almost?)

The past, however, is another story. In this new reality, our *past* is changing and evolving, as well.

What have we done?

Is the past only a stubbornly persistent illusion?

Or, as the song says, two out of three ain't bad?

Strangely enough, returning to our universe for a moment, the real Stephen Hawking wrote something on just this subject, something that could easily exist in both realities.

In *The Grand Design*, Stephen Hawking (and Leonard Mlodinow) describe a typical scientific experiment (the two-slit experiment, which, we learn, has actually been done with "buckyballs,"—soccer-ball-shaped molecules made up of sixty carbon atoms). This experiment illustrates how quantum particles take every path simultaneously to get somewhere. They then elaborate:

"This idea has important implications for our concept of "the past." In Newtonian theory, the past is assumed to exist as a definite series of events... There may have been no one watching, but the past exists as surely as if you had taken a series of snapshots of it. But a quantum buckyball cannot be said to have taken a definite path from source to screen. We might pin down a buckyball's location by observing it, but in between our observations, it takes all paths. Quantum physics tells us that no matter how thorough our observation of the present, the (unobserved) past, like the future, is indefinite and exists only as a spectrum of possibilities. The universe, according to quantum physics, has no single past, or history."

Our alternate Stephen Hawking, on the other hand, with his alternate proposal, actually manages to *restore* the past back to its original meaning, with a slight update. We will let this alternate Stephen explain:

What we used to think of as *the past*—as the fourth-dimensional past states of our evolving three-dimensional universe—we must now think of as the fifth-dimensional past states of our evolving fourth dimensional spacetime continuum. In this reality, then, our past still exists as *the* past— unchanging and unalterable—only now it is a fifth-dimensional past, rather than a fourth-dimensional one.

It is simply a matter of terminology.

Our *fourth*-dimensional past, however, *is* changing and evolving, equally with our present and our future. It is this concept that may take some getting used to—but it is certainly no more strange than quantum reality, itself.

It should also be pointed out that, in this cosmology, the past, the present and the future equally exist—just like in Einstein's relativity. Except that now, instead of being static and eternally unchanging, *all* of the modes of time are changing and evolving. It is the concept that *anything* in reality is static and unmoving that is actually the stubbornly persistent illusion. *Everything is evolving, continuously.*

A spinning fourth-dimensional spacetime continuum clarifies another issue within the no-boundary proposal. The singularity at the beginning of time, rounded off in the no-boundary proposal, is the exact center of the universal particle—and so it is not moving, fifth-dimensionally. The spin of the rest of the universal particle is the physical mechanism that causes time to "switch on" with the movement.

We humans have always looked behind us, back in time, searching for clues as to where we came from, how we came to be. The beginning of time we see as the ultimate origin of everything, mysterious and unknowable. But now we realize that "the beginning of time" is really just the center of the universal particle. It is no more mysterious than the rest of the universe.

# THE STUBBORNLY PERSISTENT

It is the rest of the universe—all of the universe—that is mysterious. Where did it come from? How did it come to be?

These questions will occupy physicists far into the future. For example, in order for the spin of the fourth-dimensional spacetime continuum to have meaning, one may ask: in relation to *what* is it spinning? Perhaps it could simply have an intrinsic spin, not dependent upon anything external—but a more natural explanation would be that it is spinning in relation to an outside cosmos—a cosmos that exists far and beyond what we perceive as our universe.

A multiverse.

It would be within this multiverse that we would then ask the questions, where did we come from? How are we here?

If we assumed that our spinning fourth-dimensional spacetime particle is not unique, then we could assume there are other spinning fourth-dimensional particles out there—other universes.

The fifth-dimensional cosmos awaits our further exploration, far into the infinite future.

Meanwhile, there are an enormous number of other physical consequences that a moving, spinning fourth-dimensional spacetime continuum will produce, all of them strange and outright bizarre. The physicists and cosmologists in this alternate universe, it turns out, have already explored the consequences of this movement for many years, working out all of the various ramifications.

They call these consequences *quantum mechanics.*

For example, they realized that all mass and energy in the universe derives from the spin of the fourth-dimensional particle. (We explore this subject in the next chapter.) For now, if we simply accept this statement as is, we can consider the consequences that result.

Because all mass and energy result from the fifth-dimensional spin of the universal particle, the *momentum* of an ordinary quantum particle arises fifth-dimensionally, by definition. (Momentum is equal to mass times velocity, which are both just alternate forms of energy.) In order to measure the momentum, then, one needs to measure the fifth-dimensional movement of the quantum particle. Despite the new terminology, this is easily done, with the appropriate physical measurement.

*However*, if one wishes to measure the *position* of a quantum particle, then one needs to measure this position within the fourth-dimensional spacetime continuum *at an instantaneous fifth-dimensional moment*. If it is not measured at an instantaneous fifth-dimensional moment, then the particle will be evolving, fifth-dimensionally, and the exact fourth-dimensional position will be lost.

# THE STUBBORNLY PERSISTENT

When one measures the exact position of a quantum particle within the fourth-dimensional spacetime continuum, at that exact instantaneous fifth-dimensional moment, the quantum particle is motionless, fifth-dimensionally, and so *it does not have a fifth-dimensional momentum.* At that instantaneous moment, its momentum has become completely, objectively indefinite.

Likewise, if one wishes to measure instead the exact momentum of a quantum particle, then one needs to measure its exact movement, fifth-dimensionally. Doing this, however, loses all information about its location within a static fourth-dimensional spacetime. One can imagine the fourth-dimensional spacetime continuum running up the page, and the fifth dimension running at a right angle, left and right on the page. One can measure the property of a quantum particle either vertically, or horizontally, but you must choose which one it is. You can never measure both the horizontal and the vertical properties *at the same time.* You can measure either the particle's fourth-dimensional properties, or its fifth-dimensional properties, but you must choose which one to measure.

In other words, you can measure *either* a particle's position, *or* its momentum, but it is impossible to measure both simultaneously. And, the more accurately you measure one of these properties, the more information is lost about the other property.

This, then, (in case you were not paying attention), explains the uncertainty principle—what is *really* going on down there.

(Oops—how did *that* happen?)

The fourth-dimensional spacetime continuum is spinning, fifth-dimensionally. As it spins, it is evolving according to the Schrödinger equation. When a measurement is made of a quantum particle, the measurement records either the instantaneous fourth-dimensional property of the particle, within fourth-dimensional spacetime, or else the instantaneous fifth-dimensional *evolution* of the particle, *as spacetime is evolving fifth-dimensionally.* Either way, the measurement is interpreted as *the collapse of the wave function.* But nothing is actually collapsing, because the fourth-dimensional spacetime continuum continues spinning—it continues to evolve according to the Schrödinger equation. The observation simply recorded an instantaneous fifth-dimensional moment, and in the process disturbed the ongoing evolution.

Most of the mystery of quantum mechanics revolves around the phrase, *at the same time.* For example, you cannot measure a quantum particle's position and its momentum *at the same time.* A particle travels from point A to point B by taking every possible path *at the same time.*

# THE STUBBORNLY PERSISTENT

A quantum particle exists in many different states *at the same time.*

You can either measure a quantum particle for particle-like properties, or for wave-like properties, but you can never measure both of these properties at the same time.

What has now become evident is that the phrase *at the same time* is actually referring to two *different* times. The first time is the one everyone normally assumes—time within the fourth-dimensional spacetime continuum. But now another type of time has emerged—a fifth-dimensional time. This is the time of the fourth-dimensional spacetime continuum as it is *moving* through the fifth dimension. What has been confusing is that this movement is taking place at what we always thought of as a single moment— say, two o'clock on a Monday morning. Even though it 'remains' two o'clock within the fourth-dimensional spacetime continuum, this moment is *itself* changing and evolving, fifth-dimensionally.

When a particle takes every possible path simultaneously, then, it is actually taking them at different fifth-dimensional moments, even though it remains the 'same' moment, fourth-dimensionally. When a particle simultaneously has many different states, it possessed these states on previous fifth-dimensional moments, which we see as the 'same' moment, fourth-dimensionally—what we observe as a superposition of states. When we make an observation, we are capturing its current, instantaneous fifth-dimensional state.

The inescapable conclusion is that there are *two time dimensions.* There is a fourth-dimensional time, and a fifth-dimensional time.

If you design an experiment to measure something in fourth-dimensional time, you get a fourth-dimensional measurement. If you design an experiment to measure something in fifth-dimensional time, you get a fifth-dimensional measurement.

Or, as it is commonly phrased, the reality you observe depends upon what you choose to measure.

Quantum mechanics originated with the discovery of Plank's constant. Plank's constant is the source of all the discreteness of quantum mechanics. Plank (and Einstein, and eventually others) discovered that energy—and all other quantum properties—only came in discrete amounts. You either had this amount, or you had nothing at all—there was nothing in between. It is this property that separates the discreteness of quantum physics from the continuous spacetime of Einstein's relativity.

Plank's constant is measured in units of time and energy, which are the units of *action.* Plank's constant, then, is a measure of the smallest unit of action.

If the fourth dimension is what we normally think of as *time*, and the fifth dimensional spin is the source of *energy*, then Plank's constant, being the smallest unit of measurement of time and energy, will be defining the *instantaneous* fifth-dimensional movement of fourth-dimensional spacetime—the smallest unit of action.

And this would account for all of the discreteness of quantum mechanics.

Quantum physicists generally consider the *reduced Plank's constant* to be even more useful than Plank's constant.  This is Plank's constant divided by two times pi.  The reduced Plank's constant— called *h-bar*—is the smallest unit of *angular momentum*.  In this form, the constant defines the smallest amount of angular momentum a particle can possess.  (Technically, the smallest unit is half this constant, but that is simply an historical technicality, and does not affect our discussion.)

A spinning fourth-dimensional spacetime continuum—a spinning universal particle—will possess an intrinsic angular momentum.  All objects existing within this universe—making up this particle—will share in the angular momentum.  This, then, provides a natural, physical explanation for the spin of quantum particles.

The uncertainty principle originated with a

puzzling discovery. While attempting to figure out how position and momentum measurements fit together in the new emerging quantum mechanics, it was realized that the *order* in which they were measured mattered. Specifically, position times momentum did not equal momentum times position—subtracting these two quantities from each other did not equal zero. Instead, it equaled the reduced plank's constant times the imaginary number, *i*. What was this? It turned out to be, somewhat rearranged, *the uncertainty principle*, the defining characteristic of quantum weirdness.

If the fourth-dimensional spacetime continuum is spinning fifth-dimensionally, then the order in which these dimensions are measured would matter. Measuring the fourth dimensional position first, then the fifth-dimensional movement, would produce a different result than measuring the fifth-dimensional movement first, and then the fourth-dimensional position. The difference would be the smallest unit of angular momentum—the reduced Plank's constant, in the imaginary direction of time.

Returning now to our universe, while keeping these concepts from our alternate universe in mind, an interesting experiment you (the reader) can make is to reread some or all of your books on quantum mechanics (or simply reread some of the quotes we have offered here, in our book), and decide for yourself if the above interpretations make any sense.

Go on—give it a try. We'll wait.

In the meanwhile, we offer a few more random examples of descriptions of quantum reality. Decide for yourself if our alternate version of explanations "make sense" or not.

Here are quotes from Peter Coveney and Roger Highfield's book, *The Arrow of Time*:

"Just as we cannot know simultaneously the position and momentum of a sub-atomic particle with any precision, so there is a limitation to the accuracy with which we can measure energy within a given interval of time. The principle connects these two quantities in the same way as it links the mutual uncertainties in position and momentum. A precise measurement of an atom's energy in a particular quantum state can only be performed at the expense of uncertainty about the time it spends in that state, in other words its lifetime. But if its lifetime is known very precisely then its energy cannot be known with any certainty...

"When the wavefunction collapse occurs, all the many possibilities reduce to a single real event. This removes the symmetry between the past state of the system (potentiality) and the present state (actuality). Indeed, if one tries to use the method to retrodict the past from a given measurement result, one gets incorrect results. Thus the very act of measurement introduces an arrow of time into the phenomena described by quantum mechanics...

"The majority of working physicists simply accept the additional postulate proposed by the Hungarian-born mathematician John von Neumann, that the wavefunction collapses on observation. No mechanism for the collapse is proposed. Indeed, it is clear that the collapse cannot be described by the Schrödinger equation itself, for that equation is reversible and deterministic, while the collapse is irreversible and random."

In Steven Weinberg's book, *Dreams of a Final Theory*, he tells us,

"By around 1930 the discussions at Bohr's institute had led to an orthodox "Copenhagen" formulation of quantum mechanics...Whether a system consists of one or many particles, its state at any moment is described by the list of numbers corresponding to each possible configuration of the system. The same state may be described by giving the values of the wave function for configurations that are characterized in various different ways—for instance, by the positions of all the particles of the system, or by the momenta of all the particles of the system, or in various other ways, though not by the position *and* the momenta of all the particles.

"The essence of the Copenhagen interpretation is a sharp separation between the system itself and the apparatus used to measure its configuration. As Max Born had emphasized, during the times between measurements the values of the wave function evolve in a perfectly continuous and deterministic way, dictated by some generalized version of the Schrödinger equation. While this is going on, the system cannot be said to be in any definite configuration. If we measure the configuration of the system (e.g., by measuring all the particles' positions *or* all their momenta, but not both), the system jumps into a state that is definitely in one configuration or another, with probabilities given by the squares of the values of the wave function for these configurations just before the measurement."

In his book, *The Cosmic Code*, Heinz Pagels warns,

"What actually happens to the electrons as they approach the barrier cannot be visualized. If you try to visualize what happens to an electron as it approaches the holes and try to figure out what happens you will, as Feynman remarked, go "down the drain." If we try to visualize the electron as a little bullet, then we should get the bullet pattern. But we don't. If we try to imagine the electron as some kind of wave, then we should detect waves on the screen. But we don't—we detect individual particles...

"I would like to emphasize that what is at issue here is the nature of physical reality. There is no meaning to the objective existence of an electron at some point in space, for example at one of the two holes, independent of actual observation. The electron seems to spring into existence as a real object only when we observe it! We cannot sensibly speak about which hole it goes through unless we set up an apparatus to actually detect it. Quantum reality is rational but not visualizable.

"...The quantum weirdness lies in the realization that as long as you are not actually detecting an electron, its behavior is that of a wave of probability. The moment you look at the electron it is a particle. But as soon as you are not looking it behaves like a wave again. That is rather weird, and no ordinary idea of objectivity can accommodate it."

Before moving on, let us return to the age-old enigma of time. From our new perspective, is it possible we might answer some, or all, of the ancient mysteries about time?

Let us find out.

First, there is the most fundamental question of them all: *what is time?* Assuming the answer must include both fourth-dimensional time and fifth-dimensional time, our new answer is: *time is the fifth-dimensional movement of fourth-dimensional spacetime.*

This has the fortunate outcome of being a *physical* definition of time, as opposed to being some vague, mystical wave of the hand.

Then there is the question, *why is time different from space?* Why, for example, can we move in any direction of space, but we can only move in one direction of time—forward—and we have *no choice* in the movement?

In our alternate cosmology, the hierarchy of dimensions easily explains the differences between time and space. Space is defined to be three-dimensional, spacetime is defined to be fourth-dimensional, and the *movement* of spacetime is fifth-dimensional.

But, why can we only move forward through time, and why do we *always* move forward through time, whether we want to or not? And what of the other, similar questions, such as: *Does time flow?* If so, *why does it flow? How does it flow?* Specifically, Why does time flow from the past to the future, and not from the future to the past? Why, in other words, is there an arrow to time?

The spin of the universal particle is the answer to all of these questions. The fifth-dimensional spin *is* the flow of time. With the spin, everything flows outwards, away from the center of the particle: energy, vibrations, centrifugal forces, causes and effects. The spin creates the arrow of time, the flow of time, and the fundamental difference between time and space—why you can only move forward through time, and why you have no choice in the movement.

*What happened before the beginning of time? Where did we come from? How are we here?*

We have already touched on some of these questions, with the realization that they are looking in the wrong direction for their answers. What we see as the beginning of time is really just the center of the universal fourth-dimensional particle, and there are no answers there. The true mystery is the existence of the particle itself—*the entire fourth-dimensional spacetime continuum*—and the answers to the mysteries of its existence reside within the vast and presently unknown fifth-dimensional cosmic universe.

The proposal that the fourth-dimensional spacetime continuum is spinning, fifth-dimensionally, of course opens up a large number of other puzzles, questions and problems. The physicists in our alternate universe have been busy working out their various solutions, with varying degrees of success. We are now going to focus on a few particular consequences of this theory, which have relevance to our own cosmology and physics.

# CHAPTER NINE

# Spinning Spacetime

In our alternate universe—which we now should perhaps call a *past fifth-dimensional state* of our fourth-dimensional spacetime continuum, or, perhaps even better, a *future* fifth-dimensional state of our fourth-dimensional universe, or perhaps even an *imaginary* future state of our past alternate universe—hey, for now on we are going to simplify our terminology, and call our real universe Cosmos I, and that other universe Cosmos II.

In Cosmos II, they believe their fourth-dimensional spacetime continuum is spinning about the center of their universal particle. Here in Cosmos I, we have a superficially similar proposal, one that has been around for a while, in which the entire observable universe is spinning. The first thing we need to make clear is that these two proposals *are not* the same. They are very different creatures.

# THE STUBBORNLY PERSISTENT

In J. Richard Gott's book, *Time Travel in Einstein's Universe*, he tells us:

"Ever since Einstein announced his equations of gravitation in 1915, people have been exploring "solutions" to them... Many of these solutions have remarkable properties. One of the most amazing was found in 1949 by Einstein's brilliant colleague at the Institute for Advanced Study at Princeton, mathematician Kurt Gödel. The solution allowed time travel to the past.

"Gödel's remarkable solution to Einstein's equations was a universe that was neither expanding nor contracting but instead rotating. Now put aside thinking about the universe for a moment, and consider yourself. Your inner ear tells you whether you are spinning or not—if you are rotating rapidly, you will get dizzy...

"Back to Gödel's universe. In that universe, a nondizzy, and therefore non-rotating, observer would see the whole universe spinning around her. From this, she could conclude that the universe was rotating. Furthermore, the distances between galaxies in Gödel's universe do not change with time...

"Yet our observations tell us that we apparently do not live in the universe proposed by Gödel. We observe that galaxies are moving away from each other—the universe is expanding. With all the orbits of the planets, asteroids, and comets, the solar system constitutes a giant gyroscope, and we can determine that the distant galaxies are not rotating relative to it. Also, if the universe had a significant amount of rotation, the temperature of the cosmic microwave background would vary in a systematic way over the sky—something we don't observe."

In Gödel's rotating universe, it is the visible universe that is rotating. But in the universe of Cosmos II, the entire fourth-dimensional spacetime continuum is spinning about the singular point at the 'beginning of time'—the center of the universal particle. The difference between these two proposals is profound.

To picture a Gödel rotating universe, imagine our ice cream cone. If you hold the ice cream cone as it should be held—bottom down, large end up—the bottom is the beginning of time, and the top of the cone is 'the moment now.' We are in the center of the layer at the top of the cone. If we hold the cone in one place, and spin the cone counterclockwise, the observable universe appears to be spinning around us—much as it would if we were in the center of a tornado. But no matter where we are in the universe, it will appear as if we are at the center of the cone, and the entire universe is spinning around us.

This is *not* how they picture their spinning spacetime in Cosmos II. In this cosmology, the *entire* ice cream cone would be rotating about the small end of the cone—the bottom tip—so that the cone is going 'up and down' (rather than left to right). This picture seems a little strange, however—out of balance, with the big end revolving around the little end—so a better analogy would be picturing a spinning three-dimensional ball, with the center of the ball representing the beginning of time—the center of the particle—and the outer ball representing 'the moment now.' In this picture, Gödel's rotating universe would be at one of the 'poles' of the ball, so that the rest of the universe appears to be spinning around while the observer appears to be in the center of the rotation. But in Cosmos II the situation is more like the equator of the ball, where all of space is spinning the same amount—so that no one notices, and no one gets dizzy. Again, this is only an analogy, allowing us to picture in three dimensions what is actually a fourth- and fifth-dimensional phenomena.

The Gödel rotating universe does not agree with the phenomena we observe in our universe. The spinning spacetime of Cosmos II, however, describes phenomena that very accurately agrees with what we observe in our universe.

The physicists in Cosmos II call the description of this phenomena *general relativity.*

Let us return to one of the basic questions of

physics.  What is gravity?

Gravity is the curvature of spacetime.  The curvature of spacetime is what we perceive as the force of gravity.  These statements are usually accepted "as is"— we ourselves have done this— but let us now go further into these concepts, and find out what they actually mean.

*Mass is the source of gravity.*  It is *mass* that causes spacetime to curve.

So, what is mass?

In Ian Sample's book, *Massive*, we found out that:

"Newton said mass was a quantity of matter that arose from an object's volume and density.  An object's mass governed its inertia, or how much it resisted being pushed around, and also how strongly it felt the force of gravity.  With these definitions in place, Newton derived the basic laws of motion…

"Einstein showed that mass and energy are interchangeable, that mass can be considered a measure of how much energy an object contains…

"Between them, Newton and Einstein laid the foundations of our understanding of the nature of mass, but in the 1960s it was clear that something was still missing.  Scientists could not explain where the fundamental particles got their masses from.  It was this mystery that Higgs's theory seemed to solve.  It gave scientists their best hope yet of fully describing the mass of everything they knew...

"Though it wasn't clear at the time, Higgs's theory pointed to a critical moment in the birth of the universe. In the immediate aftermath of the Big Bang, the cataclysmic explosion that flung the universe into existence, the elementary particles were entirely massless. Then, a fraction of a second after the Big Bang, something happened: an energy field that permeated the fledgling universe switched on. Massless particles that had been zipping around at the speed of light were caught in the field and became massive. The more strongly they felt the effects of the field, the more massive they became."

All of the above has some relevance to our story; however, there is one particular aspect of mass that Ian Sample only briefly touches upon, so let us delve deeper into this subject.

In Isaac Asimov's science book, *Understanding Physics: Light, Magnetism, and Electricity,* we learn:

"The Newtonian view of mass had dealt, really, with two kinds of mass. By Newton's second law of motion, mass was defined through the inertia associated with a body. This is "inertial mass." Mass may also be defined by the strength of the gravitational field to which it gives rise. This is "gravitational mass." Ever since Newton, it had been supposed that the two masses were really completely identical, but there had seemed no way of proving it. Einstein did not try to prove it; he merely assumed that inertial mass and gravitational mass were identical and went on from there."

Einstein, in fact, began formulating his general theory of relativity with a single overriding concept, called *the equivalency principle*. This principle states that the effects of gravity are equivalent to the effects of acceleration. For example, he would imagine a disc, spinning in space. The spinning disc would create certain effects, due to the acceleration of the spinning disc, and Einstein assumed that gravity produced those same effects. He used this type of reasoning extensively.

He also would imagine a windowless box in space, a box connected by wires to (what we will call) a spaceship, and would ask the question, if objects fall to the floor of the box, is this because the spaceship is pulling upwards on the box, or is it because the spaceship is holding the box above a planet, and the planet's gravitational field is pulling down on the box? The two explanations, he realized, were equivalent.

The principle of equivalence directly relates to the two distinct masses, because inertial mass is *defined* by the force needed to *accelerate* the mass. Gravitational mass, by contrast, is defined by the gravitational force the mass produces.

We have, then, two distinct types of mass defined by two distinct types of force. So why are they equal to one another, in the same body? Why is gravitational mass equal to inertial mass?

In their book, *Inside Relativity*, Delo E. Mook and Thomas Vargish discuss this puzzle:

"Now, despite the similarity of names, *gravitational* mass and *inertial* mass represent two very different properties of matter. Inertial mass measures the resistance an object offers to a change in its state of motion... Gravitational mass measures the strength with which one object attracts another by the force of gravity. Newton determined (and subsequent experiments have verified) that the inertial mass value and the gravitational mass value for any given body are equal... the explanation for the identical motion of all objects acted upon by gravity is contingent upon the equality of gravitational and inertial mass in Newtonian physics. But this only makes the puzzle more profound.

"Why should two such disparate properties of an object as its gravitational mass and its inertial mass be described by the same number? This puzzle has bothered physicists since Newton's day. It certainly bothered Einstein."

Einstein, of course, "solved" this problem in his own special way. However, Mook and Vargish draw an interesting comparison between Newton's explanation of gravity, and Einstein's:

"The Newtonian "law of gravity" is really a *descriptive* model of the phenomenon, and the description is quite good. However, even Newton was bothered by his model. He wanted...to comprehend a wider domain of validity than was accessible to his model; he wanted some further "explanation " of the phenomenon of gravity.

"...Einstein's model of gravity still fails to "explain" the phenomenon of gravity in the sense desired by Newton. Just as Newton's model was unable to "explain why" masses should attract one another, Einstein is unable to "explain why" masses should affect spacetime. He can describe the effect, but he provides no reason for its dependence upon matter."

In his book, *Patterns in the Void*, the astronomer Sten F. Odenwald describes his own personal journey into the mysteries and mathematics of general relativity and gravity, until he finally realizes:

"Like a latter-day Newton, I could now study black holes, but I was no closer to understanding what gravity *really* was than Newton had been 300 years earlier. Then again, you could read all of Einstein's writings, only to discover that he didn't have any great ideas about gravity, either. He could describe how it worked, not what it was."

# THE STUBBORNLY PERSISTENT

Later in his book, Odenwald tells us,

"The mathematics of general relativity is so complex and subtle that it took over fifty years before physicists like Wheeler had explored enough of it to declare that spin doesn't exist within its logical fabric... Although general relativity claimed that matter causes space-time, there wasn't a single property of atomic matter that could be found hidden in the mathematics of the theory. So long as physicists stuck to Einstein's simple geometric model for the gravitational field, there could be no other particle in the universe than the graviton itself."

And then, a little later, Odenwald describes how,

"A number of physicists have looked at the difficult task of unifying gravity with the other forces and have concluded that perhaps they can never be joined. Richard Feynman, one of the developers of quantum electrodynamics, took an increasingly dim view of gravity unification, questioning whether the incredible difficulty of taking that last step was a clue that gravity is simply different from everything else. Perhaps it really isn't a force at all and can never be described by quantum theory. If this is the case, then gravity must be treated absolutely separately from other things in the world."

It is interesting to hear Einstein's own words

on the subjects of mass, matter and inertia. In *The Meaning of Relativity*, he tells us,

"What innovations in the post-Newtonian development of the foundations of physics have made it possible to overcome the inertial system? First of all, it was the introduction of the field concept by, and subsequent to, the theory of electromagnetism of Faraday and Maxwell, or to be more precise, the introduction of the field as an independent, not further reducible fundamental concept. As far as we are able to judge at present, the general theory of relativity can be conceived only as a field theory. It could not have developed if one had held on to the view that the real world consists of material points which move under the influence of forces acting between them. Had one tried to explain to Newton the equality of inertial and gravitational mass from the equivalence principle, he would necessarily have had to reply with the following objection: it is indeed true that relative to an accelerated coordinate system bodies experience the same accelerations as they do relative to a gravitating celestial body close to its surface. But where are, in the former case, the masses that produce the accelerations? It is clear that the theory of relativity presupposes the independence of the field concept."

In an earlier chapter of our book, we listed

some of the explorations physicists here in our universe have taken towards the "lower" dimensions, the dimensions from zero to three. General relativity is a description of the fourth dimension—it is a fourth-dimensional field theory. In physics, a field is "an independent, not further reducible fundamental concept." In general, we can loosely equate the field concept in our universe with the hierarchy theory of dimensions in Cosmos II. We will explore this further in a moment, but first we will finish up our own explorations of the concepts of gravity and mass, here in Cosmos I.

In *The Lightness of Being*, Frank Wilczek defines Einstein's "first law" as his famous equation: *energy equals mass times the speed of light squared.* He then rearranges this equation into Einstein's "second law": *mass equals energy divided by the speed of light squared.* With this definition, mass is defined by its energy. He explains:

"The concept of energy is much more central to modern physics than the concept of mass. This shows up in many ways. It is energy, not mass, that is truly conserved. It is energy that appears in our fundamental equations, such as Boltzmann's equation for statistical mechanics, Schrödinger's equation for quantum mechanics, and Einstein's equation for gravity...

"With Einstein's second law, it becomes possible to think of a good answer to the question...What is the origin of mass? It could be energy. In fact, as we'll see, it mostly is."

Throughout his book, Wilczek delivers on this goal, explaining how the internal energy of particles results in the mass of 95% of normal matter. (For example, most of the mass of the nucleus of atoms derives from the energy of the quarks inside the nucleus, as they hurl around at close to the speed of light). But he then summarizes what cannot be explained—such as the actual masses of the quarks, and the mass of the electron, because these are fundamental particles.

He continues:

"Then there's the Higgs particle, sometimes said to be "the origin of mass" or even "the God particle"...In brief, the Higgs field (which is more fundamental than the particle) enables us to implement our vision of a universal cosmic superconductor and embodies the beautiful concept of spontaneous symmetry breaking. These ideas are deep, strange, glorious, and very probably true. But they don't explain the origin of mass—let alone the origin of God. Although it's accurate to say that the Higgs field allows us to *reconcile* the existence of certain kinds of mass with details of how the weak interactions work, that's a far cry from explaining the origin of mass or why the different masses have the values they do. And as we've seen, most of the mass of normal matter has an origin that has nothing whatsoever to do with Higgs particles."

According to the physicists of Cosmos II, their entire fourth-dimensional spacetime continuum is a single particle—a fourth-dimensional particle—that is spinning about its center. We can view this particle fourth-dimensionally—as the physical particle of the fourth-dimensional field, perhaps—or fifth-dimensionally, with the fourth-dimensional particle acting as the carrier of force of the fifth-dimensional field (perhaps). Or we can analyze reality three-dimensionally, as a three-dimensional space moving fourth-dimensionally (with all the various three-dimensional field theories included). Or we could analyze its two-dimensional properties, as two-dimensional field theories, or one-dimensionally, or zero-dimensionally. The reality one discovers depends upon which level of physical reality one chooses to investigate.

Reality is a hierarchy of physical dimensions.

The entire fourth-dimensional spacetime continuum is spinning, fifth-dimensionally. The spin of the particle is the source of all forms of energy, within spacetime, in all its various guises. Physicists in Cosmos II know how to equate all the forms of energy within their universe with the kinetic energy of the spinning particle, and with its angular momentum.

All inertia, for example, is a result of the

spin. What we perceive as an object at rest, or in uniform motion, is actually an object spinning with the fourth-dimensional particle. It takes a force to alter its inertial forward movement.

The spin of the particle, however, is continuously altering the fifth-dimensional straight-line movement of the lower dimensional objects within fourth-dimensional spacetime, producing a centrifugal force on them. This centrifugal force on the particles pushes them against the fabric of spacetime, causing it to bend and warp.

Let us take this down a few dimensions, for illustrative purposes.

Imagine a spinning merry-go-round. If you stand in the center of the merry-go-round, you do not feel any acceleration. The farther out from the center you stand, the more acceleration you feel, and the harder you need to hold on so that you do not go flying off. What is happening here?

A motionless body is called an inertial frame of reference. A body moving in a straight line, at a constant speed, is also an inertial frame of reference. A body at rest will remain at rest, and a body in uniform motion will remain in motion, unless acted on by some outside force. These are the starting principles of Newtonian physics.

When you are moving on a merry-go-round, your body wants to continue forward in a straight line, but the merry-go-round (assuming you are holding on) keeps pulling you away from that straight line. This is felt as a centrifugal force. But what you feel as centrifugal force is the merry-go-round pulling you away from the straight line you would naturally travel in (because the merry-go-round is propelling you forward). The spin of the merry-go-round is exerting a force on you, changing your course (because you are holding on). This change of course is called *acceleration.* While on the spinning merry-go-round, you feel a constant acceleration, away from the straight line of inertial momentum. (This is technically defined as a *centripetal* force, towards the center of the merry-go-round.)

If you let go of the merry-go-round, the centrifugal force disappears, and you would then travel in that straight line that your body naturally wanted to travel in, at the same speed you had before. The line you travel in is at a right angle to the line that goes from you (on the merry-go-round) to the center of the merry-go-round. But from your perspective, you are traveling 'away from the center' of the merry-go-round, which is the definition of *centrifugal.*

Your inertial momentum is what is propelling you forward—the tendency of all bodies to continue forward at a uniform speed in a straight line. The merry-go-round has been pulling you away from this inertial momentum, and this is what you feel as a 'force.' This means that the centrifugal force is not really a 'force' at all—the force you feel derives from your change of inertial momentum. In other words, there is not a force "pushing" you off the merry-go-round, instead there is a force pulling you away from your natural inclination to continue forward in a straight line.

For this reason, centrifugal force is called a "fictitious" force, because it is not really a true force, at all. What you are feeling is actually the acceleration of an inertial state, and an inertial state is the tendency of all bodies to move forward at a constant speed, in a straight line. So what is felt as a centrifugal force is really the acceleration of an inertial state of motion.

In almost all science books (in Cosmos I) that describe how mass curves spacetime, there will be a picture of a ball resting on a rubber sheet. The ball presses down on the rubber sheet, causing the sheet to bend. If a marble is tossed towards the ball, the marble's path will bend around the ball, following the curvature of the rubber sheet, instead of traveling in a straight line. This illustrates how matter "bends" spacetime.

What is rarely pointed out is that, in order for the ball to bend the rubber sheet, it must be experiencing the force of gravity. Otherwise, it would just be sitting flat against the sheet—or it would actually go floating away. You need the force of gravity on the ball to bend the rubber sheet. (The marble, also, must be experiencing the force of gravity.)

In other words, the force of gravity is being used to explain the force of gravity.

Perhaps a better illustration would be to have the rubber sheet in the shape of a very large, hollow ball, floating in outer space. You then place the smaller ball (made of metal, perhaps) on the *inside* of the rubber ball, and then *spin the larger ball.* Now the smaller ball will press against the rubber sheet, as it is trying to travel in a straight path, but the rubber sheet keeps pulling it away from that straight path. In many science fiction movies they show a space station as a spinning wheel, allowing people to walk around as if they were experiencing the force of gravity. The principle is the same. The acceleration of the spinning wheel is mimicking the force of gravity. *Acceleration* is used in place of gravity—Einstein's equivalency principle at work.

Now, let us return to the larger question. *Why does mass curve spacetime?* Or, more scientifically, *how* does mass curve spacetime?

If you listen for it, you can hear the physicists in Cosmos II yelling, *the acceleration of the spinning universal particle causes mass to curve spacetime.* Mass wants to travel in a straight line— a fifth-dimensional straight line—but the spinning particle is continuously pulling it away from that straight line. Mass is being accelerated, which causes it to "push" against the fabric of spacetime. In the picture of a ball sitting on a rubber sheet, it is the inertial force of the ball that is causing the sheet to bend—the centrifugal force.

The curvature of spacetime is the result of the acceleration of an inertial body, which means that it is an acceleration that is curving spacetime—but the curvature of spacetime *is gravity.* This is the reason, then, why acceleration is equivalent to the force of gravity. The one is creating the other. In fact, the one *is* the other.

The inertial force is the result of the fifth-dimensional spin, which pushes against fourth-dimensional spacetime—which results in the curvature of fourth-dimensional spacetime—which we see as gravitational mass *and* inertial mass. When viewed fourth-dimensionally, this mass is causing fourth-dimensional spacetime to curve, as gravitational mass; when viewed fifth-dimensionally, this mass is an inertial mass that resists acceleration.

Gravitational mass is a fourth-dimensional manifestation, then, while inertial mass is a fifth-dimensional manifestation, but they are *the same mass.* This is the reason why inertial mass is equivalent to gravitational mass. They are one and the same, existing in both the fourth dimension and the fifth dimension.

In order for this to work, the lower dimensions—the dimensions from zero to three— must be woven into the spacetime fabric, so that when they experience the centrifugal force, they will push against the spacetime structure. But this is exactly how the physicists in Cosmos II view their reality— as a hierarchy of dimensions.

Mass, then, originates as one of the lower dimensions (probably all of them) pushing against the fourth dimension, causing it to warp and bend. In this scenario, gravity and mass are completely defined by the physical dimensions making up the universe. Gravitational mass is a lower dimension pushing against a higher dimension, which causes spacetime to curve, which is 'gravity.'

But if both mass *and* the curvature of spacetime are the result of the physical dimensions experiencing the inertial force of the spinning fourth-dimensional universal particle, then both mass *and* gravity are the result of a centrifugal force—*meaning gravity is a fictitious force.*

There really is no force of gravity, in the physics of Cosmos II. This would be the reason why, here in Cosmos I, physicists have had so much trouble trying to unite gravity with the other three forces of nature.

There is no force of gravity. (Or, if we wish to travel full circle, back to Einstein: The force of gravity is only a stubbornly persistent illusion.) There is only the force of the spinning universal particle, and its angular momentum. All mass and energy originate with these two aspects of a spinning spacetime.

Energy equals mass times the speed of light squared, or, for Einstein's second law, mass is equal to energy divided by the speed of light squared. The second law has an interesting form, because dividing by a speed squared puts the equation in the form of *per second per second*, which is the same form as an *acceleration.* The equation, essentially, is saying that mass is energy continuously accelerated at the speed of light. This is what mass is, essentially, as will become (somewhat) clearer in the next chapter.

There are many scientific speculations, here in Cosmos I, concerning black holes. One of the more troubling concerns is the formation of a singularity at the center of a black hole. A singularity is a point of infinite density and pressure, where the laws of physics break down. One of the good and bad aspects of general relativity is that it predicts the formation of these singularities—it predicts a point in which the theory of general relativity becomes meaningless. This is both a positive aspect—most theories do not do this, leaving it up to humans to figure out where the boundaries of a theory lie—and a negative one, because it hints that the theory, as it stands, is not complete.

In the physics of Cosmos II, there are no singularities, as we will explain in a moment. This means that general relativity is a complete description of fourth-dimensional spacetime, with no need for modification or replacement by another theory. It is a complete description of the fourth-dimensional field (with the slight modifications we discussed above).

Mass originates with the centrifugal force of the spinning fourth-dimensional universal particle. As more mass accumulates at a specific point in spacetime, the more centrifugal force spacetime experiences, and the more warping and bending will occur. At some point in this process, the amount of bending and warping will become so great that it will exceed the ability of spacetime to "hold together," and the fabric of spacetime will rupture (or, perhaps, will stretch and warp without bound). The centrifugal force will exceed the force holding the fabric of spacetime together, and the fabric of spacetime will tear. The mass and energy that has accumulated at this point will be "released" from the hold of spacetime. This mass and energy will go flying out of the particle, into the multiverse.

There will then be a "hole" in the fabric of

spacetime—a black hole. Spacetime will immediately begin attempting to repair this hole, in a process known as Hawking radiation.

From the point of view of the multiverse, our universe (our fourth-dimensional particle) has become radioactive, ejecting mass and energy into the cosmic void.

Another interesting area of scientific speculation concerns time travel. The unusual feature of a Gödel rotating universe is that it allows time travel into the past. If you were able to travel into the past, you might create a series of events in which you are never born. But if you are never born, then you cannot travel into the past. And if you never travel into the past, then you make no changes, and you end up being born. So then you do travel into the past...

This is a time paradox, the fodder of science fiction stories.

In Cosmos II, the (fourth-dimensional) past is changing as much as the present and future, so time travel presents no paradoxes. If someone did travel back in time, whatever changes they brought about would travel outward at the speed of light. But, according to special relativity, an object at rest is actually traveling through *time* at the speed of light, so the effects of the change in the past would never reach the "present," because the present is always moving into the future as fast as the changes from the past.

Time travel as a narrative of entertainment is alive and well in Cosmos II.

The statement that we are traveling through time at the speed of light might seem strange, for two completely different reasons. The first reason originates from our normal, everyday experience of the passage of time. We do not usually think of this passage as a "speed." And the speed of *light?* That seems *really* fast. But this is what relativity tells us, and it actually makes a lot sense. We already assume, in our normal daily lives, that we are traveling through time, from the past into the future. Relativity informs us that there is only *one* speed within our fourth-dimensional spacetime continuum, and that is the speed of light.

Our speed through time explains why time slows down the faster we travel through space: we are still traveling at the speed of light, but some of the direction of this speed is now through space. If, on our planet, we were traveling due north, we would travel a certain distance in a certain amount of time. But if we instead traveled a little east or west of due north, we would end up traveling the same distance in the same amount of time, but the distance we traveled due north would be less than if we had simply traveled due north. This is exactly the same principle. The more we travel in space, the less we travel through time.

As we approach the speed of light, almost all of our velocity ends up going through space, with only a little bit left over for going through time. And if we could actually reach the speed of light, *all* of our speed would be through space, with none of it left for the time direction. And so this illustrates (from special relativity's point of view) how there is only one speed through spacetime—the speed of light.

This also gives a visual (mental, if somewhat naive) picture as to *why*, in relativity theory, the distance in space is *subtracted* from the distance in time (or vice versa, depending on which convention is used), which derives the *proper time* of an object, which is the time the object actually experiences. The distance through spacetime remains the same, but the more that distance is in space, the less it will be in time.

The second problem with this description, of traveling through spacetime at the speed of light, has to do with spacetime itself. According to Minkowski's spacetime, the past, the present and the future simply exist. In what sense, then, does a speed through time make any sense?

In fact, in Gödel's rotating universe (in the physics of Cosmos I), all of time *still* simply exists. If someone traveled back in time, they remain a part of the spacetime fabric, the same as everything else.

# THE STUBBORNLY PERSISTENT

They could not "change" time, any more than anyone else could. Whatever they do in the past has "already" happened, so they could never create an event in which they were never born. The most they might accomplish is to *ensure* that they ended up being born, in a strange sort of loop. At least, this is how the situation exists in Cosmos I; in Cosmos II, as explained above, the entire spacetime continuum is in a state of flux, and there are no conceptual difficulties to time travel—although there could, of course, be *physical* difficulties that might make this impossible.

In Cosmos I, the conflict between these two descriptions—of objects traveling at a certain speed through time, and the description of all of spacetime simply existing—is a curious one. Some books describe the situation the one way, some books describe the situation the other way, and some books describe things both ways, but with hardly a mention of the discrepancies between the two. This has to leave the reader slightly confused, in the end. Which description describes the reality we live in? Are we moving through time at the speed of light, or do our past, present and future simply exist?

In Cosmos II, there is no confusion. Reality is a hierarchy of dimensions, and each dimension has a different reality. In three-dimensional reality, we are moving through time; in fourth-dimensional reality, *spacetime* is moving fifth-dimensionally. The reality you observe is the reality you are in, at the moment you are observing it.

# CHAPTER TEN

## Let There Be Light

Space, by definition, is three-dimensional. Three-dimensional objects, such as apples, people and planets, exist within three-dimensional space. But all three-dimensional objects extend in time— into the past and into the future—meaning they are, really, fourth-dimensional objects. We only perceive three-dimensional "slices" of these objects, at any one moment, because our senses (and our minds) evolved to perceive a three-dimensional reality. But our science has extended our senses, so that we can now perceive more of reality, and we can intellectually grasp a fourth-dimension—as time.

This scientifically formulated fourth dimension agrees with our human intuition of time, in which we can "sense" time by remembering the past and anticipating the future. Before relativity, we could question whether or not the past and the future "actually" existed, but if we accept relativity as "true"—which physicists generally do—then we must also accept relativity's unavoidable conclusion that *all* moments in time have an equal existence. The past, the present and the future equally exist.

All three-dimensional objects, then, are three-dimensional snapshots of fourth-dimensional objects. These fourth-dimensional objects extend continuously back through time, changing shape and form, interchanging mass and energy, ultimately joining up at the singular point at the beginning of time. Reversing this picture, this singular point spreads outwards, into the vast reaches of space and time, as a single, fourth-dimensional, universal particle.

According to the physicists of Cosmos II, this fourth-dimensional particle is spinning about its center, fifth-dimensionally. Because they are all a part of this particle, and because they are all spinning together as a single entity, all three- and fourth-dimensional objects in the universe share a unique relationship with the center of the spinning particle.

The physicists of Cosmos II describe this unique relationship as *the special theory of relativity.*

Imagine a line traveling from you, wherever

you are at the moment, back through time, all the way back to the Big Bang—the center of the universal particle. You are embedded within this particle, sharing with it its physical state. You are, in fact, spinning with the particle. Because you share this fifth-dimensional forward movement with the particle, you share, along with all other objects in the universe, a special, physical relationship with the straight line that travels from the center of the particle to you.

What kind of relationship is it?

To answer this, let us explore some of the strange aspects of the special theory of relativity a little more thoroughly.

According to Einstein, the closer an object approaches the speed of light, the slower it will be moving in time. The fourth-dimensional object is rotating its time dimension into space dimensions, which is one of the reasons why its mass appears to increase, the faster it goes (there is "more" of it in the spatial dimensions). This is also the reason it appears to be traveling more slowly through time—because it is exchanging its time dimension for its space dimensions.

The faster an object travels through space, the slower it moves through time, and if the object actually reached the speed of light, its movement through time would stop completely. According to some outside observer, the object (or person) would be motionless, frozen in time (but traveling through space at the speed of light). But, according to the special theory of relativity, no material object can ever actually reach the speed of light. It takes energy to accelerate an object, and that energy adds to the mass of the object, so that more energy is needed to accelerate the object further. As the object becomes heavier and heavier, it takes more and more energy to accelerate it, until ultimately it would take an infinite amount of energy to accelerate the object all the way to the speed of light, at which point its mass would also become infinite.

No material object, then, can ever actually reach the speed of light.

Light itself, however, is massless, and it *always* travels at the speed of light. According to the special theory of relativity, then, light does not experience any passage of time. The photons that make up light, from their point of view, travel instantaneously from one point to another—even if that other point is on the other side of the universe.

Let us return to that object (or person) traveling very close to the speed of light. They appear, to a stationary observer, to be traveling slowly through time, almost frozen in time. However, from their point of view, they are experiencing the passage of time normally. *Space*, however, for them, has experienced a Lorentz contraction, in their forward motion. Stars that once seemed to be many light years away are now seen to be much closer by.

Special relativity describes how to equate these different observers, with themselves and with the universe, using the Lorentz transformations. The Lorentz transformations describe how space and time will change with an observer's motion.

Let us delve a little deeper into the meaning of these transformations.

In his book, *Great Ideas in Physics*, Alan Lightman describes the concepts Einstein wrestled with as he formed his special theory of relativity:

"A fundamental idea in relativity is that there is no condition of absolute rest or motion. There is only relative motion. Absolute motion is motion that can be determined and measured without reference to anything outside the object in motion. We frequently say that we are traveling at some speed, say 60 kilometers per hour (about 37 miles per hour). But what we really mean is that we are moving at 60 kilometers per hour *relative to the road.* In fact, the road is attached to the earth, which is spinning on its axis, and the earth is also orbiting the sun at yet another speed. The sun, in turn, is orbiting the center of the galaxy, and so on. So at what speed are we actually traveling?

"Einstein claimed that only *relative* motion exists, that an observer moving at constant speed cannot do any experiments to discover how fast, in absolute terms, she is moving or whether she is moving at all. Fixed markers in space, against which all other motions can be measured, simply do not exist. An observer can measure her speed only *relative* to another observer or object...

"Motion is closely connected to time. Another idea of relativity is that absolute time does not exist. The relative rate of ticking of two clocks depends on their relative speed."

Einstein showed how to combine these ideas so that all inertial observers will observe the same laws of physics, and will measure the same speed of light—which are the two postulates he used when formulating his special theory of relativity.

But let us back up for a moment, and delve a little deeper into the very first, primary assumption.

What is an inertial observer? What determines an inertial frame of reference?

In *Out of My Later Years*, Einstein tells us,

"The state of motion of the co-ordinate system may not, however, be arbitrarily chosen, if the laws of mechanics are to be valid (it must be free from rotation and acceleration). A co-ordinate system which is admitted in mechanics is called an "inertial system." The state of motion of an inertial system is according to mechanics not one that is determined uniquely by nature. On the contrary, the following definition holds good:—a co-ordinate system that is moved uniformly and in a straight line relatively to an inertial system is likewise an inertial system. By the "special principle of relativity" is meant the generalization of this definition to include any natural event whatever."

In his book, In *Six Easy Pieces*, Richard Feynman describes how the concept of inertia arose:

"Galileo discovered a very remarkable fact about motion... That is the principle of *inertia*—if something is moving, with nothing touching it and completely undisturbed, it will go on forever, coasting at a uniform speed in a straight line. (*Why* does it keep on coasting? We do not know, but that is the way it is.)

"Newton modified this idea, saying that the only way to change the motion of a body is to use *force*. If the body speeds up, a force has been applied *in the direction of motion*. On the other hand, if its motion is changed to a new *direction*, a force has been applied *sideways*. Newton thus added the idea that a force is needed to change the speed *or the direction* of motion of a body. For example, if a stone is attached to a string and is whirling around in a circle, it takes a force to keep it in the circle."

However, in his book, *The Void*, Frank Close asks,

"So what is an inertial frame? Answer: it is a frame where there are no net forces acting on me. And how do I know there are no net forces? Answer, when I am at rest or in uniform motion in an inertial frame. There is an awkward circularity in this."

Similarly, in Brian Greene's book, *The Fabric of the Cosmos*, we hear,

"Newton's laws of motion are usually described as being relevant for "inertial observers," but when one looks closely at how such observers are specified, it sounds circular: inertial observers are those observers for whom Newton's laws hold. A good way to think about what's really going on is that Newton's laws draw our attention to a large and particularly useful class of observers: those whose description of motion fits completely and quantitatively within Newton's framework. By definition, these are inertial observers."

Einstein incorporated inertial frames into his special theory of relativity, and told us, in his book, *Ideas and Opinions*: "The principle of inertia and the principle of the constancy of the velocity of light are valid only with respect to an *inertial system.* The field-laws also can claim to have meaning and validity only in regard to inertial systems... The four-dimensional structure (Minkowski-space) is thought of as being the carrier of matter and of the field. Inertial spaces, with their associated times, are only privileged four-dimensional coordinate systems that are linked together by the linear Lorentz transformations."

Einstein showed how inertial observers will not agree on which events are happening at the same time, and thus overthrew the concept of

# LET THERE BE LIGHT

"simultaneous events."

In his book, *Contemporary Physics and the Limits of Knowledge*, Morton Tavel asks,

"Why can't two observers in uniform motion relative to each other agree that two things happen at the same time? Surely "happening at the same time" should not depend on one's point of view. But it does. And the reason it does is that the speed of light is the same to all inertial observers. And why is the speed of light the same to all inertial observers? That's a very good question, and one good answer is that nature just behaves that way! Einstein might say that if the speed of light weren't the same for all inertial observers, it would provide a mechanism to determine your absolute speed and to determine which observer is "really" moving. This, of course, would be unacceptable to a believer in symmetry. I could be somewhat more technical and say that the speed of light is the same to all inertial observers because Maxwell's equations are the same to all inertial observers, and it is those equations that describe the motion of light and correctly predict its speed. Of course, one could then innocently ask, "Why are Maxwell's equations the same for all inertial observers?" To which I would again respond, "Nature just behaves that way!" Einstein didn't dwell on these "why" questions; he simply accepted the invariance of the speed of light as a fundamental fact of nature and went on to demonstrate its consequences."

In his book, *The End of Time*, Julian Barbour warns us,

"It is important not to be overawed by the genius of Einstein. He did have blind spots. One was his lack of concern about the determination in practice of the distinguished frames that play such a vital role in special relativity—he simply took them for granted... Matching this lack of practical interest, we find an absence of theoretical concern. Einstein asked only what the laws of nature look like in given frames of reference. He never asked himself whether there are laws that determine the frames themselves."

Let us return to the Lorentz transformations, which mathematically "transform" one inertial frame of reference into another (regarding their measurements of time, space and mass). For example, an inertial system that considers itself at rest will view another inertial system that is moving as shorter than it would normally appear (at rest). The occupants of the moving frame will be moving "in slow motion," because their time will be slowed down in relation to the rest frame. And they will appear more massive.

The Lorentz transformations show how to translate the measurements of one inertial system into another. But let us consider, for a moment, how strange this actually is.

For example, we humans have a certain

concept of the passing of time, which is basically a straight line, from the past to the present to the future. But relativity tells us that objects moving relative to us travel through time more slowly than we do. Everyone, it seems, have *their own* arrow of time, all of them pointing in different directions. The Lorentz transformations tell us how to translate one observer's viewpoint into another, but they don't tell us how to intellectually *understand* what the equations are saying.

Here, in fact, are a few questions we might naively ponder, from our human perspective.

If photons are not traveling through time, from their point of view, why do they appear to be traveling through time, from our point of view? If they must travel through time, from our point of view, why do they only travel into our future, and not our past? In fact, if they are not traveling through time, from their point of view, then why do they not simply "disappear," from our point of view, remaining at that moment in time, as they travel solely through space?

Einstein "explained" this situation, but let us view this from a different perspective. Thinking of the universe as a fourth-dimensional particle that is moving fifth-dimensionally, as the physicists in Cosmos II do, gives us an interestingly new insight.

Imagine a line going from you, right now, all the way back through time, to the Big Bang. This line travels straight through time, to the center of the universal particle. The line outwards from the center of the particle is a straight line "through" time, to you "now." We will call the center of the universal particle Point A, and we will call the point where you are Point B.

You, and all other objects in the universe, are spinning with the universal particle. Because of this, you remain "aligned" with the line that travels from the center of the particle to you. *You are always aligned with this line, from your point of view.* Whatever your apparent motion may be, you remain aligned with this straight line, because it is the source of your forward fifth-dimensional movement.

All objects in the universe are aligned with this line, because of their fifth-dimensional momentum, and this alignment, according to the physics of Cosmos II, is the definition of an inertial frame of reference. These inertial frames may not be aligned with each other—their arrows through time appear to be pointing in different directions, within the fourth-dimensional spacetime continuum—but from their *fifth* dimensional perspective, these arrows are *always* aligned with the straight line that passes through them, from the center of the particle upwards through the fourth dimension, *as this line is spinning fifth-dimensionally*. They have to be, because they are all spinning as a single particle, fifth-dimensionally.

It is this *fifth-dimensional* alignment of an

object with the spinning straight line going from the center of the particle, to the object, that orients all inertial frames of reference so that the laws of physics remain the same in all inertial systems. This alignment keeps the objects "in a straight line" with the direction of *fourth*-dimensional time, *perpendicular to their fifth-dimensional movement*, so that all objects experience the same "flow" of time, from their point of view.

How is this accomplished?

Because they are aligned with the line traveling straight up through the fifth-dimensional spacetime continuum, light always travels at a right angle to this direction. *And this keeps all the laws of physics the same for all inertial systems.*

First, let us analyze this from a fourth-dimensional perspective.

Imagine you turn on a flashlight (a very powerful one), and point it into space. The beam of light will travel away from you, at the speed of light. From the point of view of the photons making up the light wave, they are not traveling through (fourth-dimensional) time. From the perspective of the universal particle, this means that the light wave is traveling "at a right angle" to the line that goes from you to the center of the particle. The line "up" from the center of the particle is a straight line through time, and the photons are traveling at a right angle to this line, so that, from their perspective, they are not traveling through (fourth-dimensional) time. They are only traveling through (three-dimensional) space.

The light travels to Point C. From the point of view of the photons, they have not traveled through time. In fact, the line from C to B forms a right angle to the line from B to A.

Now imagine a straight line from Point C to the center of the universal particle (Point A). Because this line is the longer line (the hypotenuse) of a right triangle, it is a longer line than the line from Point A to Point B. This means that, *from the point of view of the universal particle*, the light wave has traveled into the future, even though, from the point of view of the photons themselves, they have not traveled through time. They have traveled into the future simply because the line from A to C is longer than the line from A to B (making it farther away "in time").

Now, there is a mirror stationed at Point C, reflecting the light back towards you. The light bounces off the mirror, and *again travels at a right angle,* this time from the line that goes from Point C to Point A. The light wave, again, is not traveling through time, from its point of view. When it reaches you, however, you will now be at Point D, because the right angle at Point C is tilted (upwards) in relationship to the right angle at Point B. This means that *you* have traveled through time while the light was away. The line from Point D (your present), through Point B (your past), to Point A (the beginning of time), is a straight line, a line straight through time. And the line from Point D to Point C forms a right angle to the line from Point C to Point A.

So the photons, from their point of view, have never traveled through time, while they have traveled through time from your point of view.

This also explains why light always travels into the future, from our point of view.

This discussion brings up another interesting point. If light does not travel through time, then it does not have a fourth-dimensional extension. *It is not a fourth-dimensional object.* This suggests: 1) that electromagnetism is a *lower* dimensional phenomena (the physicists in Cosmos II like to think of it as a one-dimensional dimension in three-dimensional space that is vibrating fourth- and fifth-dimensionally, and that we interpret these two different vibrational modes as electrical and magnetic phenomena), 2) that only fourth-dimensional objects have rest mass (remembering that lower-dimensional objects *become* fourth-dimensional objects, because of the centrifugal force of the spinning spacetime particle), and 3) that fourth-dimensional objects use lower dimensions to exchange energy—or, generally, that higher dimensions use lower dimensions to transfer energy. This transference of energy takes place at a right angle to the fourth-dimensional object, within fourth-dimensional spacetime. (In Cosmos II, for example, they believe a two-dimensional electron—a particle existing within two three-dimensional hyperplanes, because of the centrifugal force—emits one-dimensional electromagnetic waves, which emit zero-dimensional photons.)

One might wonder why electromagnetic waves do not "travel through time," becoming fourth-dimensional objects, like everything else. In a sense they resemble that person on the merry-go-round who "let's go," and travels at a right angle to the line from the center of the merry-go-round to the person. Because they have let go, they do not feel the centrifugal force of the spinning fourth-dimensional particle, do not become fourth-dimensional objects (and so do not have any rest mass), and do not travel through (fourth-dimensional) time.

Einstein described the electromagnetic wave this way, from his purely fourth-dimensional perspective (from an address he delivered at the University of Leiden on May 5, 1920):

"We have something like this in the electromagnetic field. For we may picture the field to ourselves as consisting of lines of force. If we wish to interpret these lines of force to ourselves as something material in the ordinary sense, we are tempted to interpret the dynamic processes as motions of these lines of force, such that each separate line of force is tracked through the course of time. It is well known, however, that this way of regarding the electromagnetic field leads to contradictions.

"Generalising we must say this:—There may be supposed to be extended physical objects to which the idea of motion cannot be applied. They may not be thought of as consisting of particles which allow themselves to be separately tracked through time. In Minkowski's idiom this is expressed as follows:—Not every extended conformation in the four-dimensional world can be regarded as composed of world-threads."

Returning now to the physics of Cosmos II, let us assume that *all* objects view reality from the perspective of their relationship with the center of the universal particle. If they are traveling through space at any significant speed, this will tilt them, fourth-dimensionally, in relationship to other fourth-dimensional objects. From their perspective, however, they always remain aligned with the fifth-dimensional spin, that is, with the fifth-dimensional spinning particle. From this perspective, light always travels at a right angle to this fifth-dimensional line that travels *through* fourth-dimensional spacetime, as it is moving fifth-dimensionally. In order to remain aligned with this line, *the entire universe appears to change, depending upon one's speed and direction,* from their point of view, and they appear to be tilted, fourth-dimensionally, from the perspective of others. This is where the Lorentz transformations fit in.  The Lorentz transformations relate all inertial frames with their fourth- and fifth-dimensional alignments.

But because they are aligned with the fifth-dimensional line traveling straight up through the fourth dimension—the line that is traveling "straight through time," but is moving fifth-dimensionally— all inertial frames observe light to travel at a right angle to this line, so that light is not traveling through (fourth-dimensional) time, from its per-spective.  And because of this alignment, the laws of physics remain the same for all inertial frames of reference.  This means that the laws of physics are the same for all inertial frames of reference *because* the speed of light is the same for all inertial frames of reference.  And the speed of light is the same for all inertial frames of reference because of their alignment with the fifth-dimensional spinning parti-cle.

The invariant interval in spacetime, the interval that all observers agree upon, is measured from this fifth-dimensional alignment.  It measures an object's *proper time*, the time the object itself actually experiences (assuming it is a single object traveling between these two points, and not two distinct objects), so that the object experiences a straight line through time, from the past to the present to the future.  It is a combination of fourth- and fifth-dimensional time—the fourth-dimensional spacetime continuum spinning fifth-dimensionally.

It is the alignment with the universal particle that defines an objects relationship with the rest of the universe, and with all physical interactions. And it is this alignment that imposes the same speed of light for all observers, no matter what their motion.

This, then, provides a purely *physical* explanation for the two postulates of Einstein's special theory of relativity.

This also explains why nothing can travel faster than the speed of light. Light, from its perspective, is not traveling through (fourth-dimensional) time. It takes zero time for light to travel anywhere in the universe—and you cannot go faster than no time at all.

This also gives a visual image as to why, if an object could go faster than light, it would end up traveling into the past.

In general, there are three specific areas of concern with the physics of Cosmos II. (We have just finished introducing the third one.) These are Newton's constant of gravity, G; Plank's constant, $h$; and the speed of light, c. In the physics of Cosmos II, all three of these constants depend upon the present size of the universal particle—meaning they will have different values in the past.

The physicists in Cosmos II are well aware of these problems, and have come up with many interesting ways of dealing with them. One of the simplest approaches is to assume that these three constants smoothly change in unison, through time, so that the laws of physics (such as the fine structure constant) remain unchanged.

This theory is extended in a natural way with the hierarchy theory of dimensions. (We might call this theory an "alternate" version of the inflationary theory of Cosmos I.) In this scenario, the singularity at the beginning of time is thought to be a true zero-dimensional point, which first expands out as a one-dimensional line, which then expands out as a two-dimensional plane, which then expands out into three-dimensional space, which expands out into fourth-dimensional spacetime, which is *still* expanding out into the fifth-dimensional cosmic universe. With each expansion taking an arbitrarily long period of time, the "beginning of time," from our point of view, lies so far back, in time, our constants of nature appear "constant" to us.

There are more imaginative theories, however. One of them takes the concept of a universal *particle* seriously, which implies that other particles exist out there in the fifth-dimensional cosmic universe—other fourth-dimensional universes, perhaps an infinite number of them. These particles might naturally form structures—for simplicity, we could imagine that two of these particles are rotating around each other, so that the "center" of our spinning universe lies far beyond our Big Bang, in between our two universes. (This is an alternate version of our braneworld scenarios.)

These are only simplified examples of the many different theories alive and well in the physics of Cosmos II.

In fact, recently another theory has become quite popular, due to an alternate version of Brian Greene. (We have quoted from the real books of the real Brian Greene, here in our real book.) This alternate Brian Greene proposed that the fourth-dimensional spacetime continuum has *finished* expanding out, fifth-dimensionally, so that all of the fifth-dimensional cosmic universe, once again, simply, eternally exists. *However*, Greene realized that if *the entire fifth-dimensional cosmic universe was moving and vibrating within a sixth-dimensional Calabai-Yai manifold in such a way...*

We have come to the inevitable conclusion that reality is infinitely complex, and infinitely layered. In this reality, anything that is possible is ultimately compulsory, meaning that anything that can exist, does exist, somewhere, somehow. For us, this means that free will *must* exist out there, somewhere, somehow.

And this is the world where we stubbornly persistent choose to live.

# CHAPTER ELEVEN

## Land of the Gods

Let us briefly summarize what our stubbornly persistent little book has accomplished.

*It has conceptually unified general relativity with quantum mechanics.* General relativity is the description of the curvature of fourth-dimensional spacetime; quantum mechanics is the description of the fifth-dimensional movement of fourth-dimensional spacetime.

This is a *conceptual* unification, which also explains why general relativity and quantum mechanics cannot be unified *mathematically.* They are each describing a different reality—a fourth-dimensional reality and a fifth-dimensional reality.

*It explains quantum reality in a manner that makes sense to our ordinary human minds*—what is *really* going on down there. It accomplishes this with the introduction of a second dimension of time—a fifth-dimensional time.

In the process, *it answers all of the ancient mysteries about time.* Time has two dimensions, a fourth-dimensional time and a fifth-dimensional time, and fifth-dimensional time is the *movement* of fourth-dimensional spacetime. This gives time a *physical* definition.

*It explains the force of gravity.* The curvature of spacetime results from the fifth-dimensional spin of fourth-dimensional spacetime, which means that gravity is the result of a centrifugal force—a fictitious force. There is no force of gravity, which explains why no one has been able to unify gravity with the other three forces of nature.

*It explains the basis of Einstein's special theory of relativity.* Inertial frames are frames aligned with the straight line traveling from the center of the universal particle to the inertial frame, and they are so aligned because they share in the fifth-dimensional forward spin of the particle. This alignment ensures that light always travels at a right angle to this line, which means that all inertial frames will witness the same speed of light, no matter what their relative velocity, and that all inertial frames will have the same laws of physics.

We have accomplished all of the above in order to achieve our one true goal: *to have free will exist within our scientifically known universe.* We did this by setting *all* of time in motion, so that, once again, there is no true distinction between the past, the present and the future—within the fourth-dimensional spacetime continuum. But within fifth-dimensional time *fourth-dimensional spacetime* has a past, a present and a future, and these modes of time agree with our human notions of past, present and future. And we are embedded within the fabric of the fourth-dimensional spacetime continuum, sharing with it its movement through time.

In this reality, the present moment is *defined* by our conscious awareness. It is the moment of our conscious awareness, as that awareness flows through time, down the length of our fourth-dimensional beings.

If nothing else, we have at least demonstrated that all of these accomplishments are *at least possible*, as a simple thought experiment. Whether or not some or all of these postulates stand the test of time, and scientific scrutiny, remains to be seen.

Or, perhaps, their validity ultimately relies on the answer to a single question: Do *you* want to live in a world with free will?

If you are not satisfied with our answer, then perhaps, at least, it might inspire you to come up with an even bigger, better solution.

Go on—give it a try. We'll wait.

In the meanwhile, *why stop here*?

Maybe, on closing, we can explain a few other things—like the meaning of existence, and whether or not there is a God.

We stubbornly persistent really do not know when to stop.

# THE STUBBORNLY PERSISTENT

Here are some existential thoughts and questions about the flow of time, the nature of human awareness, the status of free will and the meaning of our existence.

Ask yourself:

Why are we aware? How are we aware? How does an immaterial awareness arise within a material world? Does awareness have any meaning? Does it have a purpose?

Modern humans tend to identify with their awareness, but this creates existential dilemmas. If we are our awareness, then what happens to "us" when we fall asleep? Do we simply disappear? Every night? If our mind is foggy, does that mean that *we* are foggy? Are we our mental states, or do we *have* mental states?

Are we our conscious awareness, or do we *have* a conscious awareness?

When we think about other people, and when we deal with other people, we do not imagine them to be some disembodied awareness floating somewhere near their body. No, we think of them as their body. It is their body we deal with and react to, it is their body (or face) we picture when we think of them, it is their physical human body, its health and well-being, that we wish well (or to come to harm)—while assuming that their mental awareness will follow suit.

Let us find out what happens when we adopt the same perspective for ourselves.

# LAND OF THE GODS

If we are our human body, all of these existential questions disappear. Our body *has* an awareness. When we fall asleep, our body stops producing self-awareness, so that it can attend to other things. Our body has memories (of past experiences) and anticipations (of future experiences). It is the body that is grounded in the present moment, that has a concept of the passage of time, and that gives meaning to experiences.

Many people believe we have a soul, which is perfectly fine, but it does not answer any of our questions. Are we our soul, or do we *have* a soul? What happens to the soul when we fall asleep? Does it go somewhere? If it does, why don't we remember this, if this is "who" we are? Does a soul sleep? Why would it need to sleep?

We may or may not have a soul. We do know that we are not intelligent enough to know if we do or not. Therefore, we will simply let this issue go unanswered, and return to our purely scientific discussion.

Who am I? The fact that modern humans can ask this question reveals both the sophisticated state of our human development, and the confusing complexity of our modern world. Scientifically, the answer is simple: We are our physical, human body.

The body has an awareness, of itself and of the world, so that it can respond to the changes happening around it. This is the reason and the purpose of human awareness—*to ensure the*

183

*survival and well-being of our body.   Because we are our body.*

This is the reason why our body, our brain and our awareness evolved together in the first place, as (potentially) a single holistic identity.   Our body is *who* we are.

Of course, humans are highly social beings, so that the survival and well-being of the individual ultimately includes the survival and well-being of family, friends and acquaintances, in ever-widening circles of interrelationships.

We are aware of our physical, three-dimensional bodies.   But our three-dimensional bodies are a part of our fourth-dimensional beings. Our fourth-dimensional beings exist within fourth-dimensional spacetime.   We only perceive three-dimensional moments of this being, and its world, at any one time.   But we become aware of its entire existence, eventually, from one end of its being to another.

This allows us to ask, are we our three-dimensional bodies, or are we our fourth-dimensional beings?

If we identify with our three-dimensional bodies, then we live our life, we die and we are gone—we decay and disintegrate, dispersing our elements back into the universe—dust to dust.

But, if we identify with our fourth-dimensional beings, we have an eternal existence within the fabric of spacetime.

# LAND OF THE GODS

Within fourth-dimensional spacetime, our fourth-dimensional beings have an awareness—a fourth-dimensional awareness. They are aware of themselves, and of their fourth-dimensional world. How do we know this? *Because we are that awareness.* Our awareness of our three-dimensional bodies, and our awareness of our awareness, is the awareness of our fourth-dimensional beings. During our lives, we ultimately become aware of every part of this fourth-dimensional being, from one end of its existence to another. We are the first-hand account that this fourth-dimensional being is aware, of itself and of its world.

When we look back upon our lives, our memories and experiences, we define this mental journey as "our life." When the journey is over, we assume that "we" are over, and the story has ended.

Within fourth-dimensional spacetime, however, the journey has just begun.

If we are our fourth-dimensional Being, then the life we live is only one of many—perhaps an infinite number of many. Our fourth-dimensional Beings are living and surviving within a changing spacetime continuum, and what we think of as "our life" is but a single step forward, along the way.

At least, this is how the people of Cosmos II view their reality.

They believe they are fourth-dimensional Beings, living and surviving within a changing fourth-dimensional spacetime continuum. They believe their awareness flows down the length of their Beings, from one end to another, in a smooth and continuous narrative. Their Beings become aware of their world a single three-dimensional slice at a time, so that they can react and respond in a consistent manner to the changes occurring around them.

According to Einstein's theory of relativity, we exist within spacetime, eternally, our beings frozen, static and unchanging. All of space and time simply exists.

According to the physicists of Cosmos II, the entire fourth-dimensional spacetime continuum is moving, fifth-dimensionally. It is changing and evolving. This means that our fourth-dimensional Beings need to react to these changes. They react to these changes in a flowing narrative we define as our life. They live within a flowing river of time, with the free will to change their world and their destinies—one small life at a time.

The people of Cosmos II believe that the reason they perceive reality as they do—as a three-dimensional awareness of a three-dimensional body that is moving through time—is for the survival and well-being of their fourth-dimensional Beings. This is the reason and the purpose of our awareness, and why our three-dimensional world exists as a "reality."

To rephrase and summarize: *This is why our reality exists.* And it is why *we* exist.

# LAND OF THE GODS

*They do not stop there, however.*

Every person who has ever lived, and every person alive today, and every person alive in the future, *all exist equally within the fourth-dimensional spacetime continuum.* They are all fourth-dimensional Beings. They are all aware of the world, and they are all reacting and responding to an evolving spacetime continuum.

They are all communicating with one another, responding and reacting to one another, working in concert, building cities and societies and governments and religions—as if they were a single cooperative hive—or a single organism. A single fourth-dimensional super-organism evolving within a fifth-dimensional cosmic universe.

Just as our physical reality exists as a hierarchy of dimensions, so too does our human reality—and our human identity. In three-dimensional reality, we are human bodies surviving within a three-dimensional world; in fourth-dimensional reality, we are fourth-dimensional Beings living within and responding to an evolving spacetime; in fifth-dimensional reality, our fourth-dimensional Beings are but a single cell within a fourth-dimensional super-organismic being that is evolving fifth-dimensionally.

*We do not stop there, however.*

All of this *human* activity has an eerie resemblance to the ordinary human brain. Within our human brain, millions of neurons are constantly communicating with one another, cooperating with one another, reacting and responding to one another.

Within our brain, all of this activity ultimately brings forth an emergent new phenomenon: human consciousness. Human consciousness does not exist on lower levels of reality, in much the same way that water does not exist on the atomic level—water only comes into existence when two different (atomic) *gases* combine together. Water is an emergent phenomenon that only exists on the molecular level. Similarly, human consciousness does not exist on the atomic level, or the molecular level, or even the cellular level; it only arises through the interaction and communication of millions of brain cells. It is an emergent new phenomenon.

Which allows us to ask:

Within the higher levels of reality, levels we can only dimly perceive, is there an emergent new phenomenon arising from the billions of fourth-dimensional beings, beings who are all communicating and responding to each other, all of their awareness' merging into a single, holistic supra-awareness?

What would we call this super being?

Are we three-dimensional beings capable of answering these questions?

What do think?

# REFERENCES

Refences are listed in order of appearance.

Chapter One: *This Strange World*

Einstein, Albert. *Relativity, the Special and the General Theory*, 100TH Anniversary Edition, With commentaries and background material by Hanoch Gutfreund & Jürgen Renn, 2015, Princeton University Press, Princeton, New Jersey, 169-71. For an explanation of the 1954 appendix, see page xvii.

Russell, Bertrand. *The ABC of Relativity*, Fourth revised Edition, April, 1985. Mentor, New York, 70-72.

Frank, Adam. *About Time*, 2011. Free Press, New York, 137-138.

Penrose, Roger. *Cycles of Time*, 2010. Alfred A. Knopf, New York, 80.

Greene, Brian. *The Fabric of the Cosmos*, 2004. Alfred A. Knopf, New York, 130-132.

Barbour, Julian. *The End of Time*, 1999. Oxford University Press, New York, 138, 141, 143.

Davies, Paul, and Gribbin, John. *The Matter Myth*, 1992. Simon & Schuster, New York, 82.

Deutsch, David. *The Fabric of Reality*, 1997. Penguin Press, New York, 1998, 262-263.

# REFERENCES

Chapter Two: *The Stubbornly Persistent*

Penrose, Roger. *The Emperor's New Mind*, 1991, Penguin Books, New York, 304.

Davies, Paul. *About Time*, 1996, Touchstone, New York, 258, 275.

Gleick, James. *Genius—The Life and Science of Richard Feynman*, 1992, First Vintage Books, 1993, New York, 109.

Smolin, Lee. *Time Reborn*, 2013, Houghton Mifflin Harcourt, New York, xi, xxii, xxxi.


Chapter Three: *A Stranger World*

Feynman, Richard. *The Character of Physical Law*, 1995, MIT Press, Cambridge, Massachusetts, 129.

Susskind, Leonard. *The Black Hole War*, 2008, Little, Brown and Company, New York, 92-93.

Lindley, David. *Uncertainty*, 2007, Doubleday, New York, 152, 154.

Price, Huw. *Time's Arrow and Archimedes' Point*, 1996, Oxford University Press, New York, 197-98.

Lindley, David. *Where Does the Weirdness Go?*, 1996, BasicBooks, New York, 122-23.

Smolin, Lee. *The Trouble with Physics*, 2006, Houghton Mifflin Company, New York, 6.

Greene, Brian. *The Elegant Universe*, 2000, Vintage Books, New York, 3.

# REFERENCES

Chapter Five: *Our Fifth Dimension*

Kaku, Michio. *Hyperspace*, 1994, Oxford University Press, New York, 99-100, 107.

Randall, Lisa. *Warped Passages*, 2005, Harper Perennial, New York, 41-42.

Kaku, Michio. *Parallel Worlds*, 2005, Doubleday, New York, 212, 214-15.

Oerter, Robert. *The Theory of Almost Everything*, 2006, Pi Press, New York, 95.

Stewart, Ian. *Why Beauty Is Truth*, 2007, Basic Books, New York, 252-54.

Penrose, Roger. *Fashion Faith and Fantasy in the New Physics of the Universe*, 2016, Princeton University Press, New Jersey, 33, 67.

Krauss, Lawrence. *Hiding in the Mirror*, 2005, Viking Penguin, New York, 190-91.

Barbour, Julian. *The End of Time*, 1999, Oxford University Press, New York, 143-44, 145, 166-67.

Baggott, Jim. *Farewell to Reality*, 2013, Pegasus Books, New York, 200-202.

Woit, Peter. *Not Even Wrong*, 2006, Basic Books, New York, 199-200.

Feldman, Burton. *112 Mercer Street*, 2007, Arcade Publishing, New York, 156-57.


Chapter Six: *Super (Cosmic) Positions*

Davies, Paul. From the introduction to Richard Feynman's book, *Six Easy Pieces*, 1995,

# REFERENCES

Perseus Books, Cambridge, Massachusetts, xvi-xvii.

Hooper, Dan. *Dark Cosmos*, 2007, Smithsonian Books, New York, 50-51.

Penrose, Roger. *The Road to Reality*, 2005, Alfred A. Knopf, New York, 782-84.

Penrose, Roger. *Shadows of the Mind*, 1994, Oxford University Press, New York, 311-12.

Davies, Paul. *The Mind of God*, 1993, Touchstone, New York, 217.

Gribbin, John. *Schrödinger's Kittens and the Search for Reality*, 1995, Back Bay Books, New York, 161-62.

Susskind, Leonard. *The Cosmic Landscape*, 2006, Little, Brown and Company, New York, 316, 321.

Davies, Paul. *God & the New Physics*, 1984, Touchstone, New York, 116-18.

Halpern, Paul. *Edge of the Universe*, 2012, John Wiley & Sons, Inc., New Jersey, 180-82.


Chapter Seven: *Imaginary Times*

Einstein, Albert. *Relativity: the Special and the General Theory*, 100th Anniversary Edition, 2015, Princeton University Press, Princeton, New Jersey, 68-9.

Gamow, George. *One Two Three...Infinity*, 1979, Bantam Books, New York, 72-4.

Davies, Paul. *Cosmic Jackpot*, 2007, Houghton Mifflin Company, New York, 67, 76-8.

# REFERENCES

Hawking, Stephen. *The Universe in a Nutshell*, 2001, Bantam Books, New York, 80-3.

Close, Frank. *The Void*, 2007, Oxford University Press, New York, 153, 155.

Hawking, Stephen. *Black Holes and Baby Universes and Other Essays*, 1994, Bantam Books, New York, 78-9, 82.

Chapter Eight: *Quantum Continuums*

Hawking, Stephen, and Mlodinow, Leonard. *The Grand Design*, 2010, Bantam Books, New York, 82.

Coveney, Peter, and Highfield, Roger. *The Arrow of Time*, 1990, Fawcett Columbine, New York, 126, 129.

Weinberg, Steven. *Dreams of a Final Theory*, 1992, Pantheon Books, New York, 74-5.

Pagels, Heinz. *The Cosmic Code*, 1984, Bantam Books, New York, 120-22.

Chapter Nine: *Spinning Spacetime*

Gott, J. Richard. *Time Travel in Einstein's Universe*, 2001, Houghton Mifflin Company, New York, 90-92.

Sample, Ian. *Massive*, 2010, Basic Books, New York, 3, 7, 8.

Asimov, Isaac. *Understanding Physics: Light, Magnetism, and Electricity*, 1969, Mentor, New York, 117-18. Mook, Delo E. and Vargish,

# REFERENCES

Thomas. *Inside Relativity*, 1987, Princeton University Press, New Jersey, 146, 148, 168.

Odenwald, Sten F. *Patterns in the Void*, 2002, Westview Press, Boulder, Colorado, 104, 175-76, 187.

Einstein, Albert. *The Meaning of Relativity*, 2005, Princeton University Pres, New Jersey, 140.

Wilczek, Frank. *The Lightness of Being*, 2008, Basic Books, New York, 20, 201-202.


Chapter Ten: *Let There Be Light*

Lightman, Alan. *Great Ideas in Physics*, 2000, McGraw-Hill, New York, 121-22.

Einstein, Albert. *Out of My Later Years*, 2005, Castle Books, New Jersey, 54.

Feynman, Richard P. *Six Easy Pieces*, 1995, Perseus Books, Cambridge, Massachusetts, 93.

Close, Frank. *The Void*, 2007, Oxford University Press, New York, 50.

Greene, Brian. *The Fabric of the Cosmos*, 2004, Alfred A. Knopf, New York, 514-15.

Einstein, Albert. *Ideas and Opinions*, 1954, Wings Books, New York, 371.

Tavel, Morton. *Contemporary Physics and the Limits of Knowledge*, 2002, Rutgers University Press, New Jersey, 61.

Barbour, Julian. *The End of Time*, 2000, Oxford University Press, New York, 153.

Excerpt from an address given by Albert Einstein at the University of Leiden on 5 May 1920.

round robin
- with diff. b/w
random atomic &
highly correlated
from 2017 "Last pys"

- with connection to
supercondes, laser,
vs light bulb
beam gets columnated
etc,
can columnate
ordingy #4 light
not the same,

27026317R00115

Made in the USA
Middletown, DE
18 December 2018